T0205519

Functional Processing of Delta-Sigma Bit-Stream

Djuro G. Zrilic

Functional Processing
of Delta-Sigma Bit-Stream

 Springer

Djuro G. Zrilic
ECE Department
University of New Mexico
Albuquerque, NM, USA

ISBN 978-3-030-47650-2   ISBN 978-3-030-47648-9   (eBook)
https://doi.org/10.1007/978-3-030-47648-9

© The Editor(s) (if applicable) and The Author(s), under exclusive license to Springer Nature
Switzerland AG 2020
This work is subject to copyright. All rights are reserved by the Publisher, whether the whole or part of
the material is concerned, specifically the rights of translation, reprinting, reuse of illustrations, recitation,
broadcasting, reproduction on microfilms or in any other physical way, and transmission or information
storage and retrieval, electronic adaptation, computer software, or by similar or dissimilar methodology
now known or hereafter developed.
The use of general descriptive names, registered names, trademarks, service marks, etc. in this publication
does not imply, even in the absence of a specific statement, that such names are exempt from the relevant
protective laws and regulations and therefore free for general use.
The publisher, the authors, and the editors are safe to assume that the advice and information in this book
are believed to be true and accurate at the date of publication. Neither the publisher nor the authors or the
editors give a warranty, expressed or implied, with respect to the material contained herein or for any
errors or omissions that may have been made. The publisher remains neutral with regard to jurisdictional
claims in published maps and institutional affiliations.

This Springer imprint is published by the registered company Springer Nature Switzerland AG
The registered company address is: Gewerbestrasse 11, 6330 Cham, Switzerland

IN MEMORY OF
Gojko and Mika Zrilic, my parents.
Professor Georgie Lukatela, my Ph.D. thesis
adviser at Belgrade University.
Professor Savo Leonardis, my M.Sc. thesis
adviser at Ljubljana University.
Professor Dan Ross, Los Alamos National
Laboratory, faithful friend and supporter.

Preface

During the past seven decades, many articles, books, patents, and industrial design notes dealing with aspects of delta-sigma modulation (Δ-ΣM) analog-to-digital converters (ADCs) have been published.

The history of delta-modulation ADCs started with the French patent of E. M. Deloraine et al., issued in 1946. The inventors of the delta-modulator combined feedback with oversampling to achieve an increase in resolution. In order to eliminate the disadvantages of pulse amplitude modulation (PAM), they proposed per channel encoding of analog signals with pulses of constant amplitude using a 1-bit quantizer of small step size. This type of ADC is known as linear delta modulation (LDM) or just delta modulation. However, there are several problems with this type of modulation, such as slope overload, mismatch between modulator and demodulator digital-to-analog converters (DAC), and spurious pattern noises at certain types of input signals.

A more robust 1-bit ADC was proposed in 1962 by the authors H. Inose, Y. Yasuda, and J. Murakami. In combination with feedback and oversampling, they used a noise-shaping technique in the design of ADCs. This ADC design approach is known as delta-sigma modulation. There are many types of Δ-Σ ADCs, and their applications range from the low range of the frequency spectrum to the very high range of the frequency spectrum (sampling frequency order of GHz). Δ-Σ modulators are an ideal interface for biomedical and environmental sensors. Because of its high-resolution capabilities (more than 20 bits), Δ-ΣM is the leading ADC for encoding and processing of acoustic signals.

Even though there is an enormous amount of published literature on the subject of delta modulation, there are significantly smaller numbers of publications dealing with direct processing delta-modulated bit-stream. Traditionally, the decimation technique is used to bring high bit-rate of delta modulator to the speed of DSP hardware. The first implementations of a digital-differential analyzer (DDA), which go back to the late 1940s, were based on the use of delta modulation. Direct processing of Δ-Σ bit-stream provides many benefits including lower power and silicon area consumption (no decimation), signal serialization, and simple and inexpensive circuits for mixed signal applications. The main reasons for the use of delta modulation in the design of DDAs are high accuracy (high resolution) and the high interference immunity of non-positional nature of a 1-bit stream. Consequently, the accuracy of the DDA greatly exceeds the accuracy of analog computers, although they are much slower. To simulate or implement physical processes using DDA, the individual mathematical functional units are interconnected in the same way as in the case of electronic analog computers. Thus, with current advances in semiconductor technology, it is necessary to take a forward step in the development of novel mathematical functional units.

Currently, Δ-Σ ADCs are used in audio signal processing systems, communication systems, control systems, instrumentation, biomedical and environmental sensing systems, and other systems. Thus, the development of novel functional units is necessary. The key objective of this monograph is to develop a number of functional circuits which process directly the Δ-Σ bit-stream.

This monograph is organized into eleven chapters with each chapter standing as an independent unit.

In Chap. 1, some basics on ordinary delta modulation and Δ-Σ modulation are compiled from published literature. The process of transformation of LDM into Δ-ΣM is demonstrated, and the advantages of this type of ADC are highlighted. Several simulation diagrams are included as well.

In Chap. 2, Kouvaras' delta adder is introduced, and a novel contribution presenting the analytical derivation of expressions for sum and carry-out of delta adder is given. In addition, the two-point averaging circuit is introduced whose operation is based on the use of delta adder.

In Chap. 3, three novel circuits for rectification and squaring of a Δ-Σ bit-stream are introduced. Simulation results of full and half-way rectification are also presented.

In Chap. 4, two novel circuits for multiplication of two Δ-Σ modulated bit-stream are presented. The operation of both circuits is based on the same algorithm. The functionality of the proposed circuits is supported by simulation results.

In Chap. 5, a new type of RMS-to-DC converter is presented. The operation of this circuit is based on the use of the Δ-Σ rectifying encoder.

In Chap. 6, novel compressor, expander, and compander circuits are presented. Simulation results support the validity of their operation. A novel type of post-processor of a compressed signal is presented as well.

In Chap. 7, a novel system for stereo signal production is presented. This method uses a digital multiplexing technique for multiplexing left and right channels of high-resolution Δ-Σ bit-streams.

In Chap. 8, a novel type of Δ-Σ amplitude modulator is presented. The operation of the modulator is based on the nonlinear processing of Δ-Σ bit-stream in the feedback loop. The message information is contained in both the carrier and envelope of modulated signals. Simulation results support the correct operation of the proposed modulating system.

In Chap. 9, two methods for frequency deviation measurement of a source of known frequency are presented. Both methods are based on the arithmetic of Δ-Σ

bit-stream and the principles of orthogonality performed on a delta-modulated pulse stream. Simulation results show the correct operation of both proposed systems.

In Chap. 10, a novel type of voltage gain control (VGC) circuit is presented. Simulation results support the validity of the proposed circuit operation.

In Chap. 11, the Δ-Σ based integrator and differentiator circuits are presented.

Albuquerque, NM, USA   Djuro G. Zrilic

Acknowledgements

My thanks to Dr. Nikos Kouvaras of Athens whose pioneering publications on non-conventional signal processing of a delta-sigma modulated bit-stream were an inspiration to me. I am in his debt for reviewing a part of this manuscript. I am also grateful to Dr. Grozdan Petrovic who has followed, advised, and supported my research ideas over the last 40 years. Dr. Radomir Majkic of Vancouver also made valuable corrections and suggestions to improve the manuscript. Dr. Glen W. Davidson of Vanderbilt University and Mr. Joseph Stevens of Santa Fe, New Mexico, assisted with editing. Finally, I owe much to Dr. Alex Constantaras, Mrs. Helen Skinas, and Isaac Brazil for their help and friendship.

Contents

Chapter 1
Basics of Low-Pass Delta Modulation

1.1 Introduction

Multi-bit analog-to-digital converters (ADC) have dominated digital signal processing (DSP) for the last 50 years or so. There are many ADC conventional architectures such as successive approximation register (SAR) ADC, flash ADCs, integrating, ramp-compare, Wilkinson, and others. All operate at a Nyquist sampling rate f_N. Since the Nyquist rate ADC converters operate at sampling frequency f_N, which is approximately two times the maximum frequency of input signal ($f_N > = 2f_{max} = 2f_B$), a high-order low-pass filter (LPF) is required to limit higher frequency components than $f_{max} = f_B$ of an input signal entering ADC. This filter is usually referred to as an anti-aliasing filter (AAF). This is a higher-order analog filter, and it is more expensive than an entire ADC. To overcome problems of an anti-aliasing filter and complexity of an n-bit ADC, linear delta modulation (LΔM) was proposed [1]. It trades off amplitude quantization of n-bit ADC for 1-bit quantization using oversampling technique. Primary use of highly oversampled LΔM is in a LΔM-to-PCM conversion. To improve signal-to-noise ratio and dynamic range, higher-order LΔM was proposed [2]. Unfortunately, LΔM systems have problems related to stability, slope overload, and accumulation of errors during transmission. To overcome these problems, Inose and Yasuda proposed an improved method of a one-bit analog-to-digital conversion of low-frequency analog signals. The paper of Inose and Yasuda can be found in the edited book of Candy and Temes [3]. Their proposed method is known as delta-sigma modulation (Δ-ΣM). This inexpensive, low-power-consuming high-resolution ADC revolutionized VLSI System-on-Chips (SoCs) design. In 1989, Schrier and Snelgrove [4] extended oversampling and noise principles of low-pass Δ-ΣM to applications at intermediate and radio frequencies (IF and RF). This method of ADC conversion is known as band-pass delta-sigma modulation (BP Δ-ΣM) [10–13]. Because LP Δ-ΣM plays a very important role in many applications, such as sensor networks, hand-held mobile devices (cellular phone, GPS, etc.), wireless networks,

© The Editor(s) (if applicable) and The Author(s), under exclusive license to
Springer Nature Switzerland AG 2020
D. Zrilic, *Functional Processing of Delta-Sigma Bit-Stream*,
https://doi.org/10.1007/978-3-030-47648-9_1

software-defined radio, etc., we will briefly describe the principles of its operations. Both techniques are based on principles of oversampling and noise shaping. Both techniques trade off amplitude quantization of n-bit ADCs for oversampling of one-bit ADCs. This trade-off is possible under the condition of meeting a certain signal-to-noise ratio (SNR) for a specific dynamic range. In the following section, we will briefly describe operation of low-pass (LP) delta modulation.

1.2   Concept of LΔM

The concept of noise shaping ADCs was introduced by Deloraine et al., in 1946. Initially this one-bit convertor was used for voice signal digitalization and transmission. It is known as linear delta modulation (LΔM), or noise shaper (or error feedback coder) [2, 5]. For now, let us consider an analog input signal $x(t)$ as an input to LΔM. This signal has derivatives of all orders t. Using Taylor series for this signal, we can express $x(t + T_s)$ as

$$x(t+T_s) = x(t) + T_s x'(t) + \left((T_s)^2/2!\right)x''(t) + \ldots \approx x(t) + T_s x'(t), \text{for small} \, T_s \quad (1.1)$$

This equation shows that from knowledge of the signal and its derivatives at instant t, we can predict a future signal at $(t + T_s)$. Let us write the Eq. (1.1) in discrete form

$$x(k+1) \approx x(k) + T_s\left[x(k)-x(k-1)\right]/T_s \approx 2x(k)-x(k-1) \quad (1.2)$$

This equation shows that we can find a crude prediction of $(k + 1)$-th sample from two previous samples. Approximation of (1.1) improves as we add more terms in the series on the right-hand side. The oversampled LΔM system, which uses a binary quantizer, is sometimes called a differential pulse-code modulation (DPCM) system. Transmitter (encoder) and receiver (decoder) are shown in Fig. 1.1. We can see that linear delta modulation requires two integrators. The output of the integrator in the feedback loop tries to predict the input $x(k)$. Thus, the integrator works as a predictor. The output of the modulator is a quantized error (the difference between input and feedback signal). This error signal, $\Delta_q(k)$, integrates in the receiver and produces an identical signal as the modulator's feedback integrator.

From Fig. 1.1 we obtain

$$X_q(k) = X_q(k-1) + \Delta_q(k) \quad (1.3)$$

or $X_q(k - 1) = X_q(k - 2) + \Delta_q(k - 1)$
Substituting this equation into (1.3) we obtain

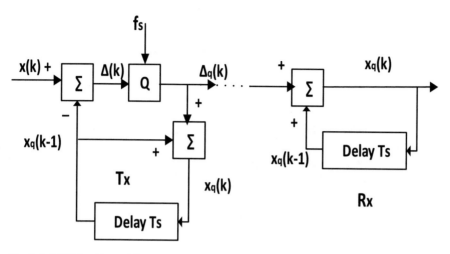

Fig. 1.1 DPCM or linear delta modulation system

$$X_q\left(k\right) = X_q\left(k-2\right) + \Delta_q\left(k-1\right) + \Delta_q\left(k\right)$$

If we continue iterating and assume a zero-initial condition for all k less or equal to 0, we can write

$$X_q\left(k\right) = \sum_{i=0}^{k}\Delta_q\left(i\right) \tag{1.4}$$

It can be seen from Eq. (1.4), and Fig. 1.1 as well, that the decoder (demodulator, receiver) is an accumulating adder. Thus, we may replace both accumulator, in Fig. 1.1 with an integrator. From Eq. (1.3) we can see that the modulated signal, $\Delta_q(k) = X_q(k) - X_q(k-1)$, carries information about the difference between successive samples. This difference is called delta (Δ), and hence, delta modulation. If the difference $\Delta(k)$ is positive, then the positive pulse is generated at output of LΔM and vice versa. We can conclude that a LΔM pulse stream carries information about the derivative of input signal $x(t)$. Because both integrators in the feedback and at the receiving (demodulating) end are linear networks, we get linear delta modulation (LΔM). Initially, LΔM was used in speech processing and digital data transmission. Unfortunately, LΔM has two main downfalls: slope overload and sensitivity to channel errors. For deeper analyses of these problems, we direct readers to references [2, 5, 6]. We will explain both effects briefly because of their importance for understanding delta-sigma modulation (Δ-ΣM), which overcomes these problems.

- First, consider a sketch below.

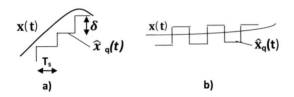

a)   b)

If $x(t)$ changes too fast, then the feedback signal $\hat{x}_q(t)$ is not able to follow changes of $x(t)$ (sketch a). If an amplitude of $x(t)$ is smaller than the step value δ of the integrator in feedback, then $x(t)$ is not possible to recover (sketch b). To avoid slope overload and have correct tracking of $x(t)$ by $\hat{x}_q(t)$, the following condition must be satisfied:

$$\left| x'(t) \right| < \frac{\delta}{Ts} = \delta f_s, \qquad (1.5)$$

where δ is the height of the step of feedback integrator, and f_s is the sampling frequency. If the input signal is $x(t) = A\sin\omega t$, then the condition for no overload is $\left| x'(t) \right|_{t=0} = \omega A < \delta f_s$. Thus, the maximum overload-free amplitude of input signal $x(t)$ is

$$A_{max} < \delta f_s / \omega. \qquad (1.6)$$

One can see that the maximum overload-free amplitude of the input signal is inversely proportional to the frequency of the input signal.

When LΔM is used in transmission, channel errors can have catastrophic effects. This is because both the integrator in the feedback of the transmitter and the receiving integrator are identical, and they must describe the same amplitude state at their outputs. When channel error happens the receiving integrator exhibits at its output a different amplitude state then the feedback integrator. In particular, this is critical in transmission of low amplitude DC like telemetry signals.

To illustrate both phenomena, we simulated the LΔM system shown in Fig. 1.2 using the MATLAB Simulink block diagram subroutines.

In Fig. 1.3, relevant waveforms are shown when the LΔM system is correctly oversampled. Signal 1 is the input signal $x(t)$, and its reconstruction version $x_q(t)$ is signal 2. Signal 3 is a difference signal $\Delta(k)$ (error signal). Signal 4 is a demodulated signal $X^\wedge(t)$, and signal 5 is a digital bit-stream, which is low-pass filtered to obtain signal 4. The error signal can be minimized by increasing the sampling frequency. In Fig. 1.4, relevant waveforms are shown when the system is overloaded. Overloading occurs when amplitude or frequency of the input signal increases. We can see an increase in amplitude of the error signal. In Fig. 1.5, influences of channel errors are shown. We can see an increase in amplitude of the error signal.

In all Figs. 1.3, 1.4, and 1.5, y and x axes present normalized amplitude and normalized time.

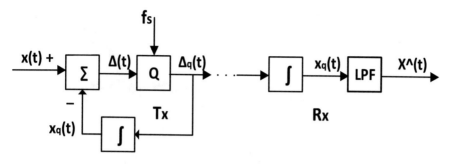

Fig. 1.2 LΔM simulation block diagram

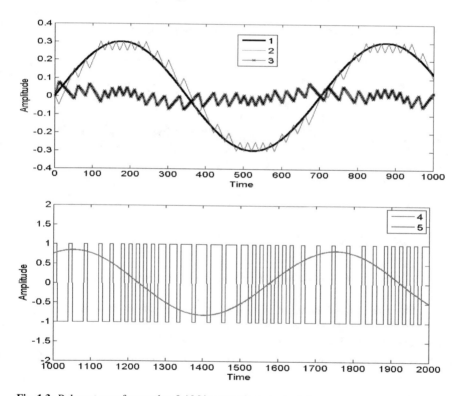

Fig. 1.3 Relevant waveforms when LΔM is correctly oversampled

1.3   Concept of Δ-ΣM

To avoid problems of slope overload distortion, accumulation of channel errors, and
instability when a higher-order filter is used in feedback of the LΔM transmitter,
Inose and Yasuda [3] proposed the concept of delta-sigma modulation (Δ-ΣM). The
idea of Δ-ΣM is based on the transformation process of LΔM into Δ-ΣM shown in
Fig. 1.6.

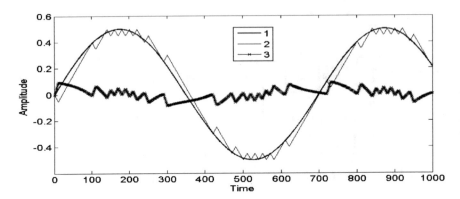

Fig. 1.4 Relevant waveforms showing effect of slope overload of LΔM

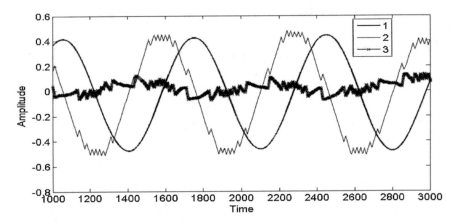

Fig. 1.5 Relevant waveforms showing influence of channel errors in LΔM system

In Fig. 1.6a, a block diagram of an LDM system is shown. To avoid aliasing, a simple filter (integrator) is placed in front of LΔM, Fig. 1.6b. As a consequence, the differentiator placed on the receiving side. Because integration and differentiation cancel, LPF becomes demodulator (receiver). Following notation in Fig. 1.6 we can write

$$\Delta(k) = X(t) - X_q(t) = \int s(t)\,dt - \int \Delta_q(k)\,dt = \int \left[s(t) - \Delta_q(k) \right] dt. \quad (1.7)$$

From Fig. 1.6c, we see that signal $X_q(t)$ becomes $\Delta_q(k)$. Thus, the error signal $\Delta(k)$ is the sum (difference) of input and quantized output signal. Binary quantizer Q encodes this sum (integral) into a one-bit signal, and because of an integration, the delta-sigma modulation system is insensitive to the rate of change of the input signal.

This transformation is possible because addition is a linear operation, and because the integrator and differentiator are linear operators. Thus, two integrators

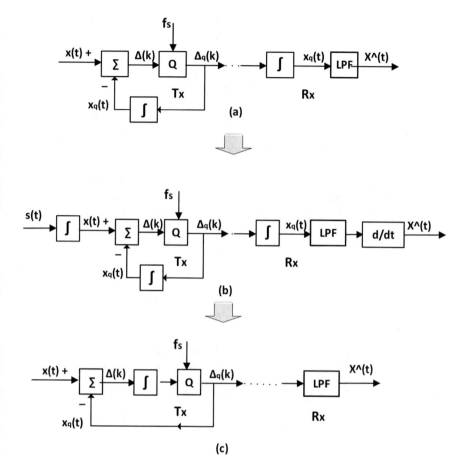

Fig. 1.6 Transformation process of LΔM into Δ-ΣM

are replaced by one integrator, Fig. 1.6b. The system in Fig. 1.6c was originally named delta-sigma modulation by its inventors. The name delta (Δ) comes from the differential amplifier output, which produces difference Δ(k), and the name sigma comes from the integrator ($\int \equiv \Sigma$) placed in the direct path of the modulator. The name Δ-ΣM corresponds to the order of mathematical operations performed in Fig. 1.6c, first difference (delta) and then integration (sigma). In this book we will keep the original name Δ-ΣM proposed by its inventors [3]. A more elaborated explanation can be found in [4].

In practice, the Δ-Σ demodulator is a moving average LPF. Thus, the influence of channel errors is not as critical as in the case of a LΔM transmission system. Let us see which maximum input signal amplitude can tolerate a delta-sigma modulator and what its dependency is on an input signal frequency. For a signal $s(t) = A\cos\omega t$, input to Δ-ΣM in Fig. 1.6b is $x(t) = \int s(t)dt = (A/2\pi f)\sin\omega t$. The maximum slope of signal is

$$\left| x'(t) \right|_{t=0} = A_{max}.$$

On the other hand, from (1.5), the maximum slope is δf_s. Thus, to avoid slope overload, the following relation must be satisfied:

$$A_{max} < \frac{\delta}{Ts} = \delta f_s. \tag{1.8}$$

We see that the maximum allowed amplitude of the input signal for Δ-ΣM does not depend on the frequency of the input signal as in the case of LΔM. Higher-order LΔM, because of a higher-order integrator in feedback, may cause serious instability problems. This is not the case with higher-order Δ-ΣM because integration is performed in the direct path of an error signal. For example, Texas Instruments Δ-Σ converters include second- through sixth-order modulators [7]. In Fig. 1.7, relevant waveforms of a Δ-ΣM system are shown. Signal 1 is an input signal $x(t)$. Signal 3 is a digital output of quantizer $\Delta_q(k)$. Signal 2 is the output of differential amplifier $\Delta(k)$. Signal 4 is an output of demodulator (LPF) $x^\wedge(t)$.

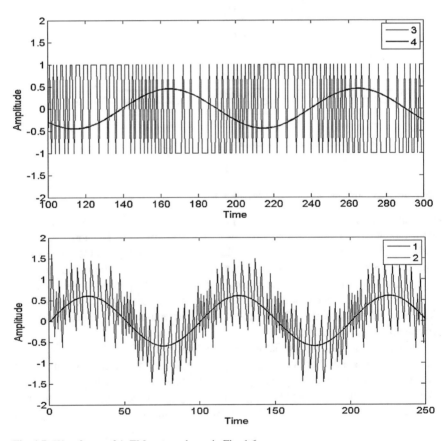

Fig. 1.7 Waveforms of Δ-ΣM system shown in Fig. 1.6c

1.3.1 Fundamental Parameters of Δ-ΣM System

As mentioned in the introduction, the complexity of anti-aliasing filter (AAF) is one of the problems facing implementation of parallel ADCs. In addition to simple AAF, Δ-ΣM relies on noise shaping and oversampling signal processing techniques. Here we will briefly introduce these techniques which are very important for minimization of quantization noise. There is a tutorial overview paper published by J. de la Rosa [8] with an extensive literature survey. There is a book of collected papers, edited by J. Candy and G. Temes [3] which we highly recommend to potential readers. The second edition of the book, "Understanding Delta-Sigma Data Converters," introduces novel approaches and design techniques for delta-sigma converters and is highly recommended as well [4].

A. *AAF and Oversampling*

The anti-aliasing filter (AAF) for the n-bit ADCs needs to be flat through passband and fully attenuate the signal at the Nyquist frequency ($f_N >= 2f_{max}$). This requirement is costly. The advantage of oversampled Δ-ΣM is that it simplifies the requirements placed on AAF. An oversampled ADC moves the sampling frequency f_s much farther away from the Nyquist frequency, and relaxes requirements on analog AAF. Figure 1.8 illustrates this.

The almost brick-wall Nyquist rate AAF may be very complex and expensive. This filter, using oversampling technique as shown in Fig. 1.8b, can be replaced by a simple LP filter.

B. *Quantization Error*

As can be seen from Fig. 1.6c, two basic elements that constitute Δ-ΣM are binary quantizer and noise shaper (integrator), in addition to differential amplifier.

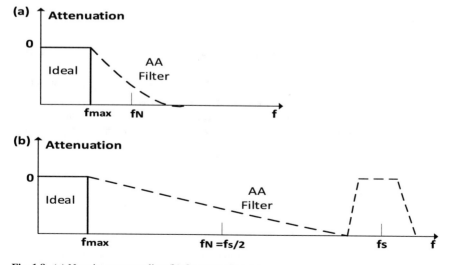

Fig. 1.8 (a) Nyquist rate sampling (b) Oversampling rate

The error introduced during the quantization process is irreversible. It is the price paid for digitalization of the analog signal. Fortunately, it can be minimized by increasing the number of quantization levels or increasing a sampling frequency (oversampling) in serial one-bit ADC. This is the price we pay for using an oversampled type of ADC. Here, in brief, we introduce what is needed for understanding direct processing of a delta modulated bit-stream. Readers interested in more information should refer to [3, 4, 8]. In our brief analysis, we assume a linear model of binary quantizer which is highly oversampled, but overload never occurs, and that the quantization error is distributed uniformly in the range $[-\delta/2, \delta/2]$ with rectangular probability density [8] shown in Fig. 1.9.

Assuming that quantization noise is white, we can write that probability spectral density in the range $[-f_s/2, f_s/2]$ as

$$S_\delta(f) = 1/f_{ss}\left[1/\delta \int_{-\frac{\delta}{2}}^{\frac{\delta}{2}} \Delta_q^2 d\Delta_q\right] = \delta^2/12fs,$$ (1.9)

and in-band noise power for low-pass signals as

$$P_\delta \equiv \int_{-f_{max}}^{f_{max}} S_\delta(f)df = \delta^2/12\left(\frac{2f_{max}}{fs}\right).$$ (1.10)

If we define oversampling ratio as

$$1/R = \frac{2f_{max}}{fs}$$ (1.11)

We can write (1.10) as

$$P_\delta = \delta^2/12R$$ (1.12)

We can conclude that with an increase of oversampling factor R by a factor of two, in-band non-shaped noise power decreases by 3 dB. This means that increasing oversampling ratio R, spectral components of quantization noise are spread thinner over a wider frequency band. This fact relaxes the shaping filter requirement. When a proper

Fig. 1.9 Probability density function of quantization noise

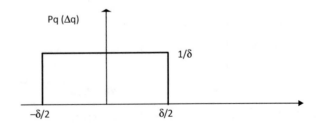

$Pq\,(\Delta q)$

$1/\delta$

$-\delta/2$ $\delta/2$

shaping filter is chosen, noise power can be reduced an additional 6 dB with doubling the sampling frequency [4].

C. *Noise Shaping*

In the case of low-pass Δ-ΣM, low-frequency in-band noise components can be attenuated using different design strategies. Assuming a linear model of quantizer, transfer function of shaping filter is defined using an S or Z-domain [3, 8]. This transfer function consists of a signal transfer function (STF) and a noise transfer function (NTF). Let us find in-band noise power for a frequency band from $-f_b$ to f_b, where $f_b = f_{max}$ of an input signal. Let us see how a high-pass filter (differentiator) operates as a quantization error reducer. Its Z-domain transfer function is given by

$$\mathrm{NSF}(z) = \left(1 - z^{-1}\right)^L \tag{1.13}$$

where L denotes the filter order [8, 9].

Assuming that $R \gg 1$, and white additive in-band noise, then the in-band shaped noise power is given by the approximate expression

$$P_q \equiv \int_{-fb}^{fb} \frac{\delta^2}{12\,fs}\left[\mathrm{NTF}9f\right]^2 df \cong \frac{\delta^2}{12} \frac{\pi^{2L}}{(2L+1)R^{2L+1}} \tag{1.14}$$

Power of this noise decreases with R by approximately 6L dB/octave [8]. Using formulas derived in reference [8], it is possible to calculate (for $L = 6$ and $R = 100$) that the signal-to-noise ratio (SNR) and dynamic range (DR) are approximately 110 dB. Let us derive signal and noise transfer function using a linear model in Fig. 1.10, assuming that quantization noise is a white additive noise.

Applying Z-transform and principle superposition for linear model in Fig. 1.10 we have

- $N(z) = 0$, then $Y(z) = H(z)[X(z) - z^{-1}Y(z)]$

$$Y(z)/X(z) = G_1(z) = H(z)/\left[1 + z^{-1}H(z)\right] \equiv \mathrm{STF}(z) \tag{1.15}$$

- $X(z) = 0$, then $Y(z) = N(z) - z^{-1}Y(z)H(z)$

Fig. 1.10 Linear model of Δ-ΣM

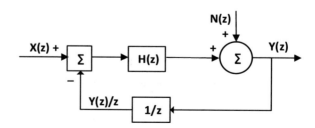

$$Y(z)/N(z) = G_2(z) = 1/[1+z^{-1}H(z)] \equiv \text{NTF}(z) \qquad (1.16)$$

Overall response of the system, due to the action of signal and noise, is

$$Y(z) = H(z)X(Z)/[1+z^{-1}H(z)] + N(z)/[1+z^{-1}H(z)] \qquad (1.17)$$

We can conclude that, for an ideal integrator, $|H(f)| \to \infty$, then STF(f) → 1 and NTF(f) → 0. This means that the input signal is allowed to pass whereas the quantization error is removed by an ideal high-pass filter. An ideal condition is difficult to meet in practice, because of a limited filter gain. For the first-order modulator discussed, the base-band noise cannot reach −98 dB signal-to-noise ratio needed for 16-bit resolution. Oversampling alone does not solve the problem. The solution is in the increase of the order of $H(z)$. There are number references, dealing with higher-order Δ-ΣM, surveyed in references [3, 4, 8]. Because our interest is limited only to the first, second, and third-order Δ-ΣM, we will briefly introduce basic concept of higher-order Δ-ΣM.

1.4   Concept of Higher-Order Δ-ΣM

When multiple first-order Δ-ΣM loops are fed-forward cascaded to obtain a higher-order Δ-ΣM system, the signal that is passed to the successive loop is the error signal from the preceding loop. In Fig. 1.11 a block diagram of Lth order feed-forward Δ-ΣM is shown.

The output signal of Lth order Δ-ΣM is

$$Y(z) = X(z) + (1-Z^{-1})^L Q(z) \qquad (1.18)$$

where $Q_L(z)$ is quantization noise from Lth order Δ-ΣM [8].

Thus, for $L = 2$, the output of second-order Δ-ΣM is

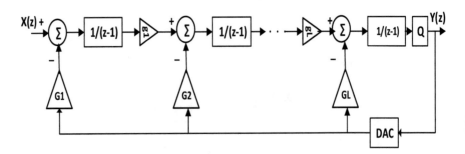

Fig. 1.11 Lth order feed-forward ΔΣM system [3]

$$Y(z) = X(z) + (1 - Z^{-1})^2 Q(z) \tag{1.19}$$

Again, using a linear model of second-order Δ-ΣM, we can derive STF(z) and NTF(z). This model is shown in Fig. 1.12.

Applying the principle of superposition to the linear circuit in Fig. 1.12 we can write:

1. $N(z) = 0$

$$\{[X(z) - Y(z)]H(z) - gY(z)\} H(z) = Y(z)$$

$$Y(z)/X(z) = H^2(z)/[1 + H^2(z) + gH(z)] = \text{STF}(z) \tag{1.20}$$

2. $X(z) = 0$

$$[-Y(z)H(z) - gY(z)]H(z) + N(z) = Y(z)$$

$$Y(z)/N(z) = 1/[1 + H^2(z) + gH(z)] = \text{NTF}(z) \tag{1.21}$$

We can conclude from (1.20) and (1.21) that, when $H(z) \to \infty$, $Y(z) = X(z)$, and $N(z) = 0$, respectively. The coefficient **g** can be chosen for optimum performance of Δ-ΣM. Higher-order Δ-ΣM pushes the quantization noise to even higher frequencies than the first- or second-order modulator. The sketch in Fig. 1.13 illustrates a noise-shaping characteristic of higher-order filters. We can see that the higher-order Δ-ΣM has less base-band noise when higher-order shaping filters are employed. Quantization error strongly depends on an input signal $x(t)$. As stated earlier, under some assumptions, which are normally met in practice, the linear model is justified. Under the same input signal and sampling frequency condition as in Fig. 1.7, relevant waveforms for second-order Δ-ΣM are shown in Fig. 1.14. We can see that the level of quantization error of the second-order system (signal 2 from above) is significantly lower than in the first-order Δ-ΣM system.

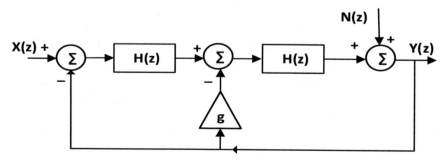

Fig. 1.12 Linear model of second-order Δ-ΣM

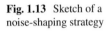

Fig. 1.13 Sketch of a
noise-shaping strategy

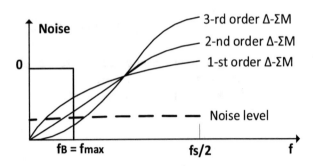

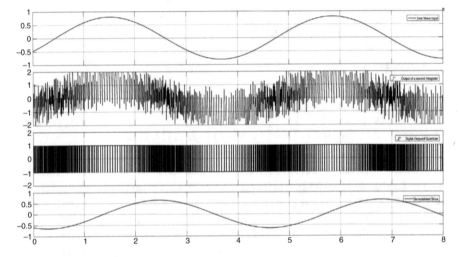

Fig. 1.14 Relevant waveforms of second-order Δ-ΣM

1.5 Summary

The objective of this chapter is to introduce the basics of operation of noise-shaping
modulators, in particular LΔM and Δ-ΣM. First, LΔM is introduced and some disadvantages of this type ADC are highlighted. Then the process of transformation of
LΔM into Δ-ΣM is demonstrated, and the advantages of this type of ADC are highlighted. The combined effect of oversampling and noise shaping are pointed out
using higher-order Δ-ΣM. In summary, we can conclude:

- The noise-shaping function in a Δ-ΣM is the inverse of the function of filter $z/(z-1)$ (which is an integrator). The inverse of the integrator is the differentiator $(1 - z^{-1})$. Thus, this filter has a dual function, passing the signal and attenuating
quantization noise.
- A filter with higher gain at low frequencies is expected to provide better base-
band attenuation of quantization noise. Higher-order Δ-ΣM systems have stron-
ger attenuation at lower frequencies.

- Even though the quantization error is correlated with an input signal, in particular for the first-order $\Delta\text{-}\Sigma M$, a linear model we employed is justified in analytical evaluations of delta modulators.
- When multiple first-order $\Delta\text{-}\Sigma M$ loops are cascaded, as in Fig. 1.11, to obtain higher-order feed-forward cascade, quantization error is attenuated according to expression (1.18).

The following chapters will focus on the circuits and systems for direct processing of a $\Delta\text{-}\Sigma$ bit-stream. Thus, we assume that a $\Delta\text{-}\Sigma$ modulator is correctly oversampled, and it operates correctly. For readers interested in $\Delta\text{-}\Sigma$ modulation, we recommend the quoted references. In addition, there is a wealth of published books and papers which can be found on the Internet.

References

1. Deloraine, E. M., Van Miero, S., Derjavich, B. French Patent # 932 140.
2. Steele, R. (1975). *Delta modulation systems*. London: Pentech Press. ISBN 0-7273-0401-1.
3. Candy, J. C., & Temes, G. C. (Eds.). (1991). *Oversampling delta-sigma data converters*. Pickataway, NJ: IEEE Press. ISBN 0-87942-285-8.
4. Pavan, S., Schrier, R., & Temes, G. (2017). *Understanding delta-sigma data converters*. Hoboken, NJ: Wiley-IEEE Press. ISBN 978-1-119-25827-8.
5. van de Plassche, R. (1994). *Integrated analog-to-digital and digital-to-analog converters*. Boston: Kluwer Academic Publisher. ISBN 0-7923-9436-4.
6. Lathi, B. P. (1998). *Modern digital and Analog communication systems*. New York: Oxford University Press. ISBN 0-19-5110009-9.
7. Related Web site: dataconverter.ti.com
8. de la Rosa, J. M. (2011). Sigma-delta modulators: Tutorial Overview, design guide, and state-of-the-art survey. *IEEE Transactions on Circuits and Systems I: Regular Papers, 58*(1), 1–22.
9. del Rio, R., et al. (2006). *CMOS cascade $\Sigma\Delta$ modulators for sensors and telecom: Error analysis and practical design*. New York: Springer.
10. Jantzi, S. A., Martin, K. W., & Sedra, A. S. (1997). Quadrature band-pass $\Delta\Sigma$ modulation for digital radio. *IEEE Journal On Solid States, 32*(12), 1935–1938.
11. Rodriguez-Vasquez, A., et al. (Eds.). (2003). *CMOS telecom data converters* (pp. 379–420). Boston: Kluwer Academic Publisher.
12. Borudopoulos, G., Pnevmatikakis, A., Anastassopoulos, V., & Deliyannis, T. (2003). *Delta sigma modulators, design and applications*. London: Imperial College Press. ISBN 1-86094-369-1.
13. Park, S. (1993). *Principles of sigma-delta modulation for analog-to-digital conversion*. Phoenix, AZ: Motorola Inc.

Chapter 2
Linear Processing of a Delta-Modulated Bit-Stream

2.1 Introduction

In the last four decades modest progress has been made in the area of direct processing of a delta-modulated stream. High resolution, simplicity, and the low cost of a delta modulation encoder have been the main reasons for using this encoder in digital signal processing. Traditionally delta modulation encoded filtering is achieved by means of Nyquist rate decimators. To avoid the process of rate conversion and decimation filtering, a number of authors recommended direct processing of delta modulation pulse stream in general, and in particular a Δ-Σ modulation bit-stream. However, because of the high oversampling rate of a Δ-Σ bit-stream, specific arithmetic circuits such as a delta adder, multiplier, etc. must be developed. Here in brief, we will mention some of the most relevant works related to linear processing of delta-modulated bit-stream.

In 1973, Peled and Liu [1] proposed a new realization of non-recursive digital filters using ordinary linear delta modulation as a means for analog-to-digital conversion. Implementation of FIR filters is based on the use of a read-only-memory (ROM). In 1974, Engle and Steenart [2] recommended implementation of a digital adder, but this adder cannot be used for implementation of a general-purpose multiplier used in realization of FIR filters. Significant advancement in hardware implementation of one-bit delta-sigma processing units was made by Greek authors, mainly Kouvaras [3–7]. Between 1978 and 1985, Kouvaras proposed several digital circuits such as delta adder, delta multiplier, delta doubler, and a number of modular networks for reduction of quantization noise. In 1978, Lagoyannis proposed a new method for multiplying delta-modulated signals by a constant [8]. In 1981, Lagoyannis and Pekmestzi proposed multipliers of two delta-sigma sequences. Use of a proposed multiplier is demonstrated in implementation of a parallel digital correlator [9]. In 1991, Horianopulos, Anastassopoulas, and Deliyannis recommended a design technique for hardware reduction in delta modulation FIR filters [10].

© The Editor(s) (if applicable) and The Author(s), under exclusive license to Springer Nature Switzerland AG 2020
D. Zrilic, *Functional Processing of Delta-Sigma Bit-Stream*, https://doi.org/10.1007/978-3-030-47648-9_2

A design technique for implementation of a high-order infinite impulse response (IIR) delta-sigma filter was proposed in [11]. A second-order variable IIR Δ-Σ filter is used as a kernel for the cascade realization of the higher-order filter transfer function. Padir and Franks presented a new digital filter recursive structure based on delta modulation [12]. In 1993, Johns and Lewis proposed a design technique for realization of IIR filters operating on oversampled Δ-Σ modulated signals. They eliminated multibit multipliers by re-modulating internal filter states [13].

A majority of the circuits proposed in this monograph are related to the use of delta adder (DA) proposed by Kouvaras [3]. This chapter is dedicated to the analytic derivation of the formulas for the sum and carry-out of a delta adder as defined by Kouvaras.

2.2   Kouvaras' Delta Adder

In 1978 Kouvaras [3] proposed an original method for addition of two delta-modulated pulse sequences $\{X_n\}$ and $\{Y_n\}$ by defining the algebraic sum S_n and carry-out C_n of a delta adder as

$$S_n = 1/2[X_n + Y_n] - 1/2[C_{n-1} - C_n] \tag{2.1}$$

$$C_n = X_n Y_n C_{n-1}, \tag{2.2}$$

where X_n, Y_n, S_n, C_{n-1}, and C_n take the value of $+1$ or -1. It was shown [3] that the demodulated sum of two delta-modulated (DM) sequences is approximately equal to the half-sum of sequences $\{X_n\}$ and $\{Y_n\}$. The generated error, because of carry-out propagation and quantization noise, can be reduced by an increase of sampling and decrease of a delta step size of delta modulator. Kouvaras [3] has shown that Eqs. (2.1) and (2.2) can be synthetized using conventional binary full adder with an interchanged role of sum and carry-out terminals. As shown in Fig. 2.1, delta adder (DA) terminals are now **Cn** = S and **Sn** = C.

Through theoretical analysis and experiments, Kouvaras verified the validity of his results designing digital filters based on the use of a delta adder. Since 1978, a number of papers have been published [14–17] using Kouvaras' delta adder without proof of where Eqs. (2.1) and (2.2) come from. Based on the above, we can conclude that the derivation of formula (2.1) and (2.2) was heuristic, which leads to the question: Is it possible to derive this equation analytically? In [17], Zrilic has shown, using the LaGrange polynomial, that it is possible to derive the algebraic equation for direct arithmetic operations on a multi-value delta-modulated pulse stream. However, it is not possible to apply this approach for a polar binary signal because of a singularity problem in the LaGrange polynomial.

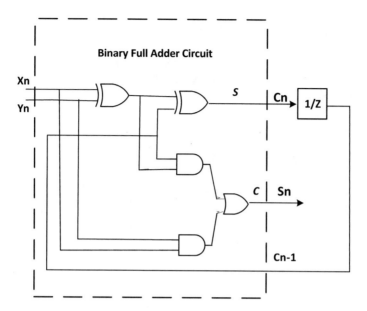

Fig. 2.1 Delta adder with output terminals **Sn** and **Cn**

Table 2.1 Truth table of delta adder

X_n	Y_n	C_{n-1}	S_n	C_n
−1	−1	−1	−1	−1
1	−1	−1	−1	1
−1	1	−1	−1	1
1	1	−1	1	−1
−1	−1	1	−1	1
1	−1	1	1	−1
−1	1	1	1	−1
1	1	1	1	1

In this section we will show how to obtain Eqs. (2.1) and (2.2) starting from the truth table of conventional binary full adder. Replacing logic "0" with "−1" and interchanging roles of the sum and carry-out terminals of a binary full adder, Table 2.1 can be constructed. Starting with this table, we will derive analytically Eqs. (2.1) and (2.2).

2.2.1   Linear Variable Dependency Approach

The elements of the truth table can be considered as a set of $\{X_n, Y_n, C_{n-1}, S_n, C_n\}$. The elements of the set are to be linearly dependent if there exist weights a, b, c, u, v, not 0, such that

$$aX_n + bY_n + cC_{n-1} + uS_n + vC_n = 0, \qquad (2.3)$$

According to Table 2.1 we can write

$$
\begin{pmatrix}
-1 & -1 & -1 & -1 & -1 \\
1 & -1 & -1 & -1 & 1 \\
-1 & 1 & -1 & -1 & -1 \\
1 & 1 & -1 & 1 & -1 \\
-1 & -1 & 1 & -1 & 1 \\
1 & -1 & 1 & 1 & -1 \\
-1 & 1 & 1 & 1 & -1 \\
1 & 1 & 1 & 1 & 1
\end{pmatrix}
\begin{pmatrix}
a \\
b \\
c \\
u \\
v
\end{pmatrix}
=
\begin{pmatrix}
0 \\
0 \\
0 \\
0 \\
0 \\
0 \\
0 \\
0
\end{pmatrix}. \qquad (2.4)
$$

Using the Gaussian method of elimination [19], we have

$$
\begin{pmatrix}
1 & 1 & 1 & 1 & 1 \\
0 & 2 & 2 & 2 & 0 \\
0 & 0 & 2 & 2 & -2 \\
0 & 0 & 0 & 2 & -4 \\
0 & 0 & 0 & 0 & 0 \\
0 & 0 & 0 & 0 & 0 \\
0 & 0 & 0 & 0 & 0 \\
0 & 0 & 0 & 0 & 0
\end{pmatrix}
\begin{pmatrix}
a \\
b \\
c \\
u \\
v
\end{pmatrix}
=
\begin{pmatrix}
0 \\
0 \\
0 \\
0 \\
0 \\
0 \\
0 \\
0
\end{pmatrix}. \qquad (2.5)
$$

From (2.5) it follows:

$$2u - 4v = 0 \rightarrow u = 2v,$$

$$c + u - v = 0 \rightarrow c = -v,$$

$$b + c + u = 0 \rightarrow b = -v,$$

$$a + b + c + u + v = 0 \rightarrow a = -v.$$

Setting $v = -1$, we have

$$X_n + Y_n + C_{n-1} - 2S_n - C_n = 0. \tag{2.6}$$

We get a similar equation as (2.1), which was defined and experimentally verified by Kouvaras [1].

$$S_n = \frac{1}{2}[X_n + Y_n] + 1/2[C_{n-1} - C_n] \tag{2.7}$$

The only difference is in the sign of the second term of Eqs. (2.1) and (2.7).

2.2.2   Algebraic Transformation Approach

From Table 2.1 we can conclude that the Boolean logic equation for carry-out is given by

$$C_n = (X_n \oplus Y_n) \oplus C_{n-1}. \tag{2.8}$$

Transforming the XOR logic truth table into algebraic [18], we obtain the algebraic truth table below. We see that modulo-2 equation, $Z_n = (X_n \oplus Y_n)$, can be replaced by the equivalent algebraic equation

$$C_n = -(-X_n Y_n)C_{n-1} = X_n Y_n C_{n-1}. \tag{2.9}$$

We can see that the algebraic Eq. (2.9) is identical to the algebraic Eq. (2.2) defined by Kouvaras [3].

X_n	Y_n	Z_n	X_n	Y_n	Z_n
0	0	0	−1	−1	−1
1	0	1	1	−1	1
0	1	1	−1	1	1
1	1	0	1	1	−1

XOR Logic truth table	**Algebraic truth table**

2.3   Δ-Σ Adder as a Two-Point Average Circuit

According to Fig. 2.2 and expression for the sum of delta adder (2.1), we have

$$S_n = \frac{1}{2}[X_n + X_{n-1}], \tag{2.10}$$

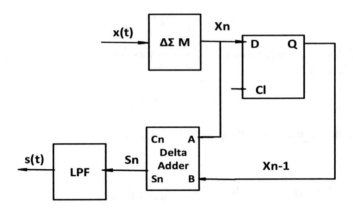

Fig. 2.2 Δ-ΣM two-point average circuit

where a term for carry-out is neglected.

Our objective is to analyze the system in Fig. 2.2. First, let us compute the transfer function $T(e^{j\Omega})$, and let us take **z**-transform of Eq. (2.10).

$$S(z) = 1/2\left[X(z) + \frac{1}{z}X(z)\right] \tag{2.11}$$

This equation leads to

$$T(z) = \frac{S(z)}{X(z)} = \frac{1}{2\left[\frac{z+1}{z}\right]}. \tag{2.12}$$

Replacing z with $(e^{j\Omega})$, we have

$$T(z)\Big|_{z=e^{j\Omega}} T(e^{j\Omega}) = 1/2\left(\frac{e^{j\Omega}+1}{e^{j\Omega}}\right) \tag{2.13}$$

$$T(e^{j\Omega}) = \frac{e^{\frac{j\Omega}{2}}}{e^{j\Omega}}\left(\frac{e^{\frac{j\Omega}{2}}+e^{\frac{-j\Omega}{2}}}{2}\right) = e^{\frac{-j\Omega}{2}}\cos\left(\frac{\Omega}{2}\right). \tag{2.14}$$

The magnitude is even and periodic in Ω with period 2π. This is a two-point **Δ-Σ** averaging filter and it will pass low frequencies ($\Omega \approx 0$) and reject high frequencies ($\Omega \approx \pi$), so it is a crude low-pass filter. The magnitude and phase of this filter are shown in Fig. 2.3.

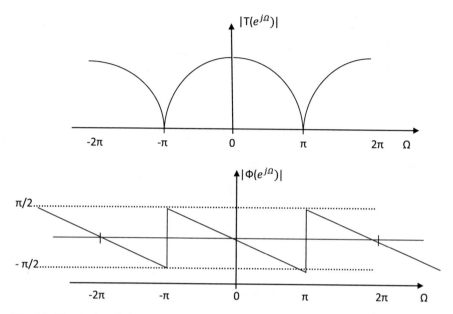

Fig. 2.3 Magnitude and phase spectra of Δ-Σ two-point averaging filter

2.4   Δ-Σ Adder as a Multiplier by a Constant

Kouvaras [3] has shown that delta adder can be successfully used to implement a multiplier by a constant less than one. A simple practical arrangement of addition/subtraction and multiplication by a constant is shown in Fig. 2.4. Let $y(t) = x(t) = ez^{-t}\sin(\omega t)$, where $f = 10$ Hz. Both modulators have the same oversampling ratio R. As shown in Fig. 2.5 the resulting signal of the sum has the value $s(t) = [x(t) + y(t)]/2$.

Using the same arrangement we see the result of addition when two analog input signals are added by means of a delta adder. In Fig. 2.6, the addition of two signals of different frequencies is shown.

Different arrangements of delta multipliers (DMPs) for multiplication of delta-modulated signal by a constant are presented in references [3, 8, 9]. One of these arrangements is shown in Fig. 2.7. A sinusoidal delta-modulated signal, $x(t) = 0.1 \sin(\omega t)$, produces bit-stream X_n. After multiplication by the constant $a = (0.5625)_{10}$ and demodulation (LPF) signal $p(t) = a\,X_n = 0.05625$ is obtained, Fig. 2.8.

Similar kinds of multiplier arrangements are used for implementation of FIR and IIR filter coefficients [3, 8, 9].

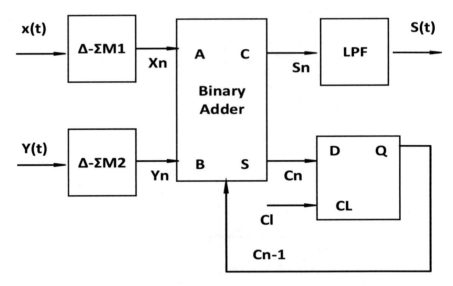

Fig. 2.4 Block diagram arrangement for addition of two Δ-Σ bit-streams

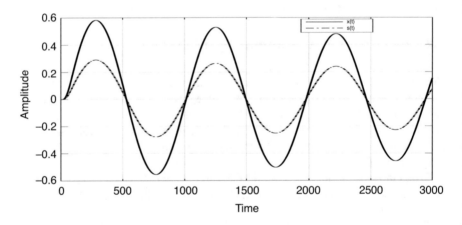

Fig. 2.5 Simulated waveforms obtained by means of the arrangement of Fig. 2.4

2.5   Summary

This chapter has briefly introduced the work of Kouvaras. His work is mainly related to linear processing of a delta-modulated bit-stream with application. Examples of implementation of FIR and IIR filters, based on direct processing of a delta-bit stream, can be found in the quoted references. The original contribution of this chapter is in the analytical derivation of formulas for the sum and carry-out defined by Kouvaras [3]. Intentionally or unintentionally, credit for invention of a delta adder was given in literature by the authors of reference [14]. However, Kouvaras

Fig. 2.6 Example of
addition of two sinusoidal
signals, $f1 = 10$ Hz and
$f2 = 15$ Hz

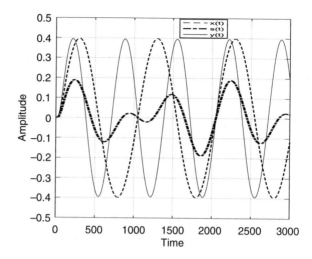

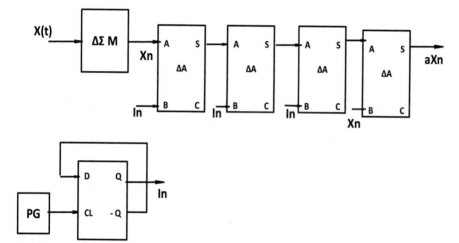

Fig. 2.7 Implementation of delta multiplier for $a = 0.5625 = (0.1111)_2$

was the first to propose a delta adder/subtractor and the first to heuristically define formulas for its sum and carry-out terminals. Thus, full credit goes to Kouvaras, who published several pioneering papers on the same subject during the period 1978–1985. A novel contribution in this chapter presents an analytic derivation of expressions for the sum and carry-out of delta adder, and a two-point averaging circuit, operation of which is based on the use of a delta adder. We can easily extend the operation of this circuit to four or n-point average circuits. Several novel circuits based on the use of a delta adder and rectifying encoder are introduced in the next chapter.

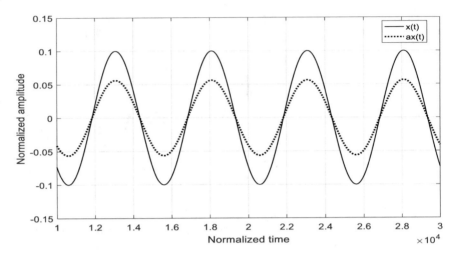

Fig. 2.8 Result of multiplication, $p(t) = ax(t)$

References

1. Peled, A., & Liu, B. (1973). A new approach to the realization of non-recursive digital filters. *IEEE Transactions on Audio and Electroacoustics, 21*(6), 477–484.
2. Engel, I., & Steenaart, W. (1976). Digital implementation of delta modulation signals. In *Proceeding of the Canadian Communication and Power Conference*, Montreal, pp. 245–248.
3. Kouvaras, N. (1978). Operations on delta-modulated signals and their application in the realization of digital filters. *The Radio and Electronic Engineer, 48*(9), 431–438.
4. Kouvaras, N. (1981). Some novel elements for delta-modulated signal processing. *The Radio and Electronic Engineer, 51*(8), 241–248.
5. Kouvaras, N., & Karakatsanis, J. (1985). A technique for a substantial reduction of the quantization noise in the direct processing of delta-modulation signals. In *Signal Processing* 8, North Holland, pp. 107–119.
6. Kouvaras, N. (1984). Novel multi-input signal-processing networks with reduced quantization noise. *International Journal of Electronics, 56*(3), 31–378.
7. Kouvaras, N. (1985). Modular network for direct complete addition of delta-modulated signals with minimum quantization noise. *International Journal of Electronics, 59*(5), 587–595.
8. Lagoyannis, D. (1878). Multiplier for delta-modulated signals. *Electronics Letters, 14*(19), 615–616.
9. Lagoyannis, D., & Pekmestzi, K. (1981). Multipliers of delta-sigma sequences. *The Radio and Electronic Engineer, 51*(6), 281–286.
10. Horianopoulos, S., Anastassopoulos, V., & Deliyannis, T. (1991). Digital technique for hardware reduction in Delta modulation FIR filters. *International Journal of Electronics, 71*(1), 93–106.
11. Pneumatikakis, A., Anastassopoulos, V., & Deliyannis, T. (1995). Realization of a high-order IIR delta sigma filter. *International Journal of Electronics, 78*(6), 1071–1089.
12. Padir, H., & Franks, L. (1988, April). A new digital recursive filter structure based on delta modulation encoding. In *Proceedings of the ICASSP*, pp. 1850–1853.
13. Johns, D., & Lewis, D. (1993). Design and analysis of delta-sigma based IIR filters. *IEEE CAS-II, 40*(4), 233–240.
14. O'Leary, P., & Maloberty, F. (1990). Bit stream adder for oversampling coded data. *Electronics Letters, 26*(20), 1708–1709.

15. Liang, L., Wang, Z. G., Meng, Q. Q., & Guo, X. D. (2010). Design of high-speed high SNR bit-stream adder based on $\Sigma\Delta$ modulation. *Electronics Letters, 46*(11), 752–753.
16. Ng, C. W., Wong, N., & Ng, T. S. (2007). Efficient FPGA implementation of bit-stream multiplier. *Electronics Letters, 43*(9), 496–497.
17. Zrilic, D. G. (2005). *Circuits and systems based on delta modulation, linear, nonlinear and mixed mode processing.* Berlin: Springer. ISBN 3-540-23751-8.
18. Petrovic, G. Personnel Communication via E-mail.
19. Majkic, R. Personnel Communication via E-mail.

Chapter 3
Rectification of a Delta-Sigma Modulated Signal

3.1 Introduction

Techniques of direct processing of the Δ-Σ modulated pulse stream emerged more than four decades ago. In 1978, N. Kouvaras [1] proposed the use of a serial binary adder as a delta adder, with an interchanged role of the Sum and Carry-out terminals of the binary adder. In that paper, Kouvaras defined the formula to find the sum of two Δ-Σ modulated pulse streams and used a delta adder as a basic processing element in the implementation of FIR and IIR filters. Over the years, a number of papers have been published dealing with linear processing of Δ-Σ modulated pulse streams [2–4]. Analytical evaluation and proof of Kouvaras' formulas were elaborated in Chap. 2.

There are also a number of publications related to the nonlinear and mix processing of Δ-Σ pulse streams. In 1995, Dias proposed a number of signal processing circuits which operate in the delta-sigma domain [5]. From 2002 to 2013 a number of Japanese authors published a significant amount of work related to nonlinear processing of Δ-Σ modulated pulse streams [6–13]. The objective of these published papers was to create a library of nonlinear circuits. The complexity of the proposed circuits hinders their applications in low-power systems. In particular, they build absolute circuits and mini/max circuits for use in median filter synthesis and in power-measurement systems [11]. Because Δ-Σ modulation is a very well-established ADC in communication receivers, bio-medical instrumentation, sensor interface circuits, measurement systems, etc., there is a need for simple, elementary signal processing circuits. One of those elementary circuits is a rectifier of AC signals.

Our goal is to design simple, reliable, and inexpensive absolute circuits, which will complement and enrich the existing library of nonlinear circuits. This design is based on direct processing of Δ-Σ modulated bit-stream.

© The Editor(s) (if applicable) and The Author(s), under exclusive license to
Springer Nature Switzerland AG 2020
D. Zrilic, *Functional Processing of Delta-Sigma Bit-Stream*,
https://doi.org/10.1007/978-3-030-47648-9_3

3.2 Δ-Σ Rectifying Encoder Implemented with Logic Circuits

A simple and inexpensive delta-sigma rectifying encoder (RE) was presented in reference [14]. It consists only of a D flip-flop and excusive OR gate (or XNOR), shown in Fig. 3.1. However, at lower input levels of an analog signal the proposed circuit introduces spikes in a rectified signal, Fig. 3.2.

To avoid spike appearance, Zrilic [16] proposed a novel RE circuit shown in Fig. 3.3.

It consists of a RE and polarity (sign) detector for elimination of undesired spikes in a rectified signal, which appear at extremely low input signal levels. The relevant waveforms of the proposed circuit are shown in Fig. 3.4. We can see a clean rectified signal which has the same input level as the signal in Fig. 3.2. At a higher input signal level ($A > 0.1$), both the RE and the RE with polarity logic detector, operate correctly without spike appearances. In Fig. 3.5 a full-wave rectified signal is shown for normalized input amplitude $A = 0.8$.

3.3 Δ-Σ Rectifying Encoder Implemented with Switch Circuits

Operation of the switch (SW) circuit is as follows. There are three inputs of the SW. Input "1"and "3" are data ports, and input "2" is a control port. For instance, data pass through input "1" when input "2" satisfies a selected criterion, otherwise, pass through input "3". In Fig. 3.6, a selected threshold is greater or equal to 0. Depending on the polarity of input signal $x(t)$, the control pulse stream C is produced.

The proposed circuit can operate as a full- or half-wave rectifier. When a Δ-Σ bit-stream O1, at output of SW1, is demodulated (low-pass filtered), then a half-way rectified signal $r(t)\_{HW}$ is obtained. This case of rectification is shown in Fig. 3.7 for a normalized input amplitude of sinusoid $A = 0.1$. When the outputs of the switches SW1 and SW2 are subtracted, and then low-pass filtered, a full-wave rectified signal $r(t)\_{FW}$ is obtained. Figure 3.8 illustrates this case. It is worth mentioning that the

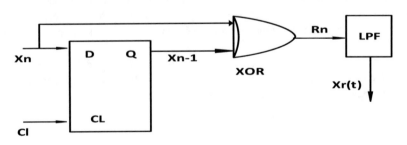

Fig. 3.1 Rectifying encoder [15]

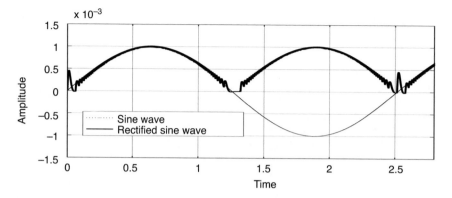

Fig. 3.2 Rectified sinusoidal signal with spikes

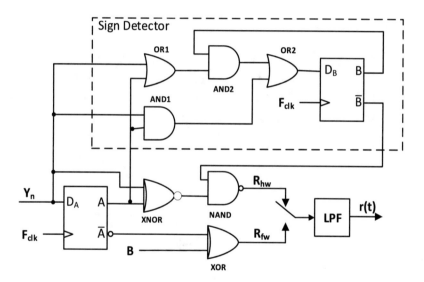

Fig. 3.3 Rectifying encoder (RE) with a polarity (sign) detector [17]

cut-off frequency of demodulating a low-pass filter is much higher than the maximum frequency of a rectifying signal (5–10 times). However, when the cut-off frequency of a demodulating filter in Fig. 3.1 and Fig. 3.3 is lower, then the proposed rectifiers operate as a squarer of a Δ-Σ bit-stream [15].

With the arrangement shown in Fig. 3.9, a squaring (nonlinear) operation on a Δ-Σ bit stream can be achieved. This figure shows the block diagram arrangement for squaring an operation using switch circuits. Figure 3.10 shows the result of a squaring for a sinusoidal input signal to a Δ-Σ modulator. One can see that an output signal $y(t)$ is DC and has a double the frequency of the input signal $x(t)$.

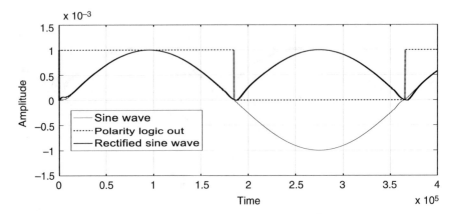

Fig. 3.4 Rectified signal without spikes

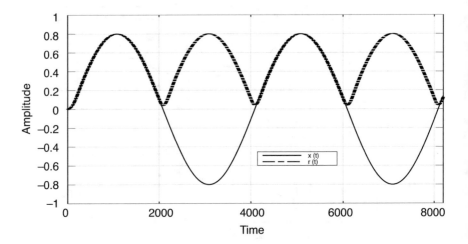

Fig. 3.5 Full-wave rectified signal for $A = 0.8$

3.4   Full- and Half-Wave Rectification of a Third-Order Δ-Σ Modulated Bit-Stream

Higher-order interpolative modulators (known as a "noise-shaping" modulators) greatly reduce the oversampling requirement for high resolution conversion applications and, in addition, randomize quantization noise to avoid the need for dithering. A novel topology for constructing stable noise-shaping modulators of arbitrary order is described in [18]. The loop stability of these modulators is determined primarily by the feed-forward coefficients. An experimental fourth-order modulator for audio application was successfully implemented and tested at MIT [18]. The objective of this paper is to employ an existing higher-order Δ-Σ modulator to show

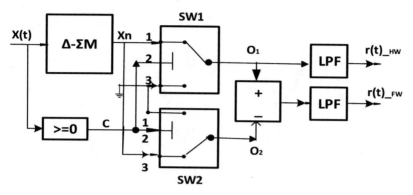

Fig. 3.6 Δ-Σ rectifying encoder implemented with switches

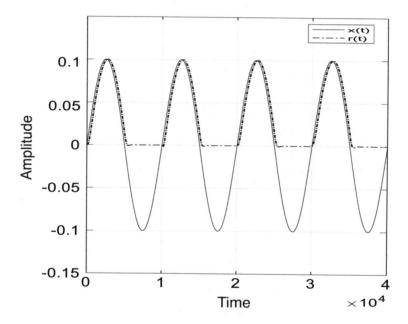

Fig. 3.7 Half-wave rectified sinusoidal signal

that full- and half-wave rectification of the third-order Δ-Σ modulated bit-stream is possible.

3.4.1   A Third-Order Δ-Σ Modulator

The transfer function of the modulator under consideration is described by an equation in the z-domain:

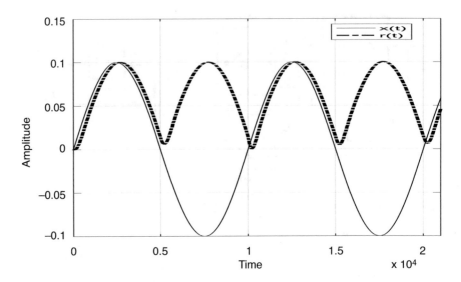

Fig. 3.8 Full-wave rectified sinusoidal signal

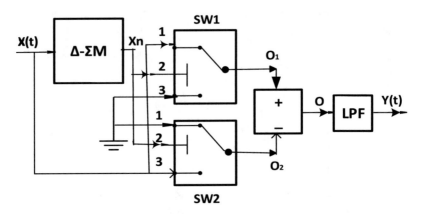

Fig. 3.9 Block diagram arrangement for squaring operation

$$Y(z) = z^{-1}X(z) + \left(1 - z^{-1}\right)^3 E(z),$$

where $(1 - z^{-1})^3 E(z)$ is the third-order modulation noise. In the analysis of bit-stream sequences we assume that the Δ-Σ modulator is highly oversampled, where the oversampling factor R is defined as $R = F_s/2F_{in}$. So, if $R \gg 1$, a change of input signal can be considered constant at the sampling instant [14]. In addition, we assume that the input signal $x(n)$ is band-limited and the spectral density function of quantization noise is uniform [8, 19]. To assure stability of the system presented in Fig. 3.11, the coefficients are chosen to be: $c_1 = b_1 = b_2 = b_3 = 1$, and $c_2 = 0.008$, $c_3 = 0.003$. In higher-order Δ-Σ modulators the in-band shaped noise decreases with

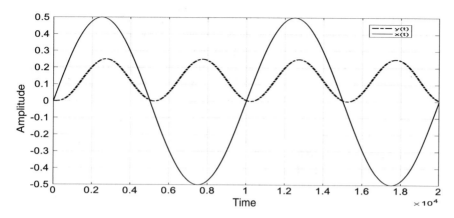

Fig. 3.10 Sinusoidal input $x(t)$, and squared output $y(t)$

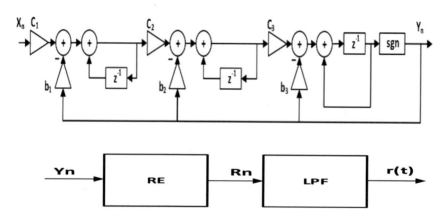

Fig. 3.11 Simulation model of the third-order Δ-Σ M rectifying system

an oversampling factor R by approximately 6 N/octave where N is the order of the shaping filter. In the case of the third order Δ-Σ modulator, $N = 3$, the decrease in noise level is 18 dB/octave. Thus, the action of a noise-shaping filter in a higher-order of Δ-Σ modulators, in combination with the oversampling, yields a significant benefit in increase of signal-to-noise ratio (SNR) [18–21].

3.4.2 Operation of Rectifying Encoder

The operation of RE with the sign detector, shown in Fig. 3.3, is as follows. A binary bit-stream Y_n of a Δ-Σ modulator (Fig. 3.11) is delayed in D_A flip-flop for one clock period to obtain signal Y_{n-1} (which is signal A). Both bit-streams are fed into AND1 and OR1 gates. The output of OR2 gate is delayed in D_B flip-flop, and its output B is fed back to the input of the AND2 gate. The output B of D_B flip-flop is XOR-ed with the delayed input bit-stream A' to obtain the rectified Δ-Σ bit-stream R_{fw}. We can see that the output of the RE circuit depends on the present states of flip-flops, D_A and D_B. Thus, it represents a Moor state machine. The transitional behavior of the proposed RE circuit is shown in Table 3.1. Keeping in mind that the next state of a D flip-flop is $Q^+ = D$, the meaning of the symbols in Fig. 3.3 and Table 3.1 is as follows:

1. Flip-flops input and output equations are
   $D_A = Y_n$; $D_B = Y_nA + B(Y_n + A)$; $R_{fW} = B$ mod2 A'
2. The next states for flip-flops are
   $A^+ = Y_n$; $B+ = Y_nA + B(Y_n + A)$; $R_{fW} = B$ mod A'
3. Corresponding maps for A^+ and B^+ are

	Yn	
AB	0	1
00	0	1
01	0	1
11	0	1
10	0	1

A^+

	Yn	
AB	0	1
00	0	0
01	0	1
11	1	1
10	0	1

B^+

Combining these two maps yields the transition table, which gives the next state of both flip-flops (A^+B^+) as a function of the present state and input, Table 3.1. The

Table 3.1 Transitional table

	A^+B^+		
AB	$Y_n = 0$	$Y_n = 1$	R_{fw}
00	00	10	1
01	00	11	0
11	01	11	1
10	00	11	0

output function R_{fW} is added to the table. Replacing 00 with S_0, 01 with S_1, 11 with S_2, and 10 with S_3 in Table 3.1 yields the state table given in Table 3.2.

The state graph of Fig. 3.12 represents Table 3.2. Each node of the graph represents a state of the RE with the sign detector, and the corresponding output is placed in the circle below the state symbol. The arc joining the two nodes is labeled with the value of Y_n which causes a state change. For example, if the circuit is in state S_0 and $Y_n = 1$, a clock edge will cause a transition to state S_3.

3.4.3 Behavioral Waveforms of a Full-Wave Rectifier

In Fig. 3.13, the behavioral waveform of a third-order Δ-Σ full-wave rectifying system is shown.

We see that full-wave rectification occurs when the output of a D_B flip-flop B is logic "1" (i.e., the negative half-period of the sine signal is flipped about the x-axis).

In Fig. 3.14, the behavioral bit-streams of the input and output of the full-wave rectifier are presented.

The top two bit-streams present inputs to the logic gate AND1 and OR1 (signals Y_n and Y_{n-1}). The third signal from the top is an output B of a D_B flip-flop (polarity sign signal). Signal B is XOR-ed with A' signal to obtain R_{fW} bit-stream. We see clearly that a width of logic "1" in the R_{fW} bit-stream has the longest duration during the picks of positive or negative half-period of the input signal. Thus, after averaging (i.e., Low Pass Filtering, LPF) the full-wave recified signal, $r(t)$ is obtained (bottom signal in Fig. 3.13). Input bit-streams are Y_n and A (which is Y_{n-1}), binary output of the polarity sign detector B (third from the top), and full-wave rectified bit-stream R_{fW} (bottom). It has been shown that rectification of the first-order Δ-Σ bit-stream is prone to the occasional appearances of spikes at low levels of the input signal. To suppress a spike appearance the RE circuit was proposed [14, 16]. In

Table 3.2 State table

PS	NS		
	$Y_n = 0$	$Y_n = 1$	R_{fw}
S_0	S_0	S_3	1
S_1	S_0	S_2	0
S_2	S_1	S_2	1
S_3	S_0	S_2	0

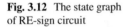

Fig. 3.12 The state graph
of RE-sign circuit

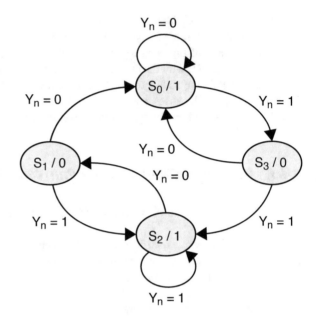

Fig. 3.15, a plot of the rectified signal without spikes for a normalized input level is $A_{in} = 0.05$. We see that the proposed RE with a sign detector completely suppresses spikes when it is used for rectification of higher-order Δ-Σ bit-streams. This was not the case with the first-order Δ-Σ rectification system, where the RE-sign circuit decreases the level of spikes only [14].

3.5   Half-Way Rectifier

Adding XNOR and NAND gates in Fig. 3.3 and switching LPF in position R_{hw} yield a half-wave rectifier. The Boolean logic equation of the rectifier is

$$R_{hw} = \text{NOT}\left[B' * \text{NOT}\left(A \bmod 2 Y_n \right) \right]$$

Figure 3.16 is a plot of the half-wave rectified signal of normalized amplitude, $A_{in} = 0.4$.

3.6   An Alternative Solution for Full- and Half-Wave
Rectifier Circuits

A slightly more complex hardware solution of a rectifying encoder is shown in Fig. 3.17. It consists of 3D flip-flops and 4 XOR gates.

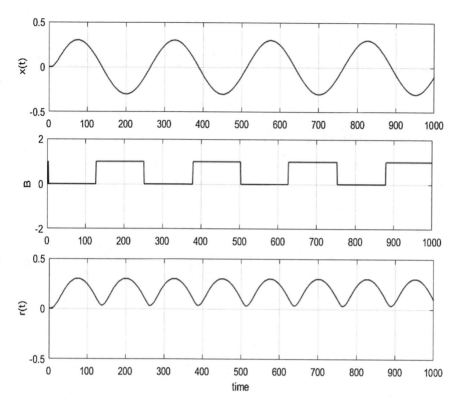

Fig. 3.13 Behavioral waveforms: analog input $x(t)$ (above), binary output B of a D_B flip-flop (sign detector output), rectified signal $r(t)$ (below)

It operates for an entire input signal level range $(-1 < x(t) < +1)$ without introducing spikes at extremely low input levels. In Fig. 3.18, rectified waveforms are seen for both rectification with and without sign detector. The normalized input signal level is $X = 0.01$ (sinusoidal signal in the middle). One can see, during zero crossing, a slight degradation of demodulated signal $x(t)^\wedge$. Thus, special attention must be devoted to the modulator and demodulator design. A proposed alternative system of rectification works well even for very low input signal levels. In Fig. 3.19 rectified waveforms are shown for the normalized input amplitude of $X = 0.003$. The upper signal presents rectification without a sign detector, and the lower signal presents a waveform when a detector is included. The middle signal presents demodulated $\Delta\text{-}\Sigma$ bit-stream. Again, we see no significant difference, except that lower signal (operation with the sign detector), and upper signal (no sign detector). As can be seen, the demodulated signal of a $\Delta\text{-}\Sigma$ modulator, $x(t)^\wedge$, is distorted. Thus, it is worth mentioning again, that special attention must be paid to the design of a $\Delta\text{-}\Sigma$ modulator regardless if a modulator is a first or higher order. Correctly behaving $\Delta\text{-}\Sigma$ is a key for correct operation of the proposed rectifiers.

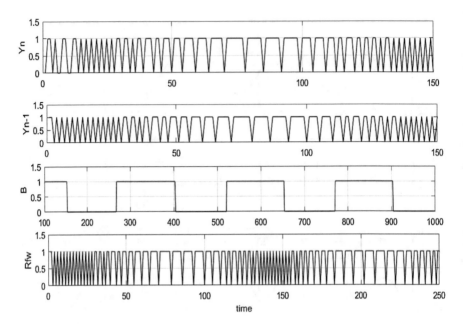

Fig. 3.14 Behavioral waveforms

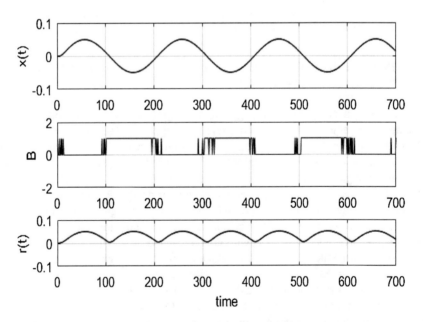

Fig. 3.15 An example of rectification when input signal is $A_{in} = 0.05$

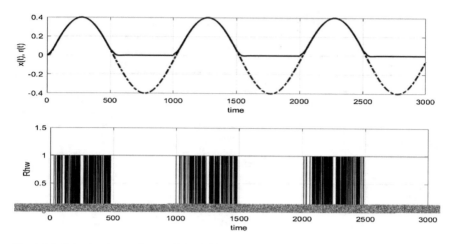

Fig. 3.16 Behavioral waveforms of the half-wave rectifier: input and rectified output signals (above), output of NAND gate, signal R_{hw} (below)

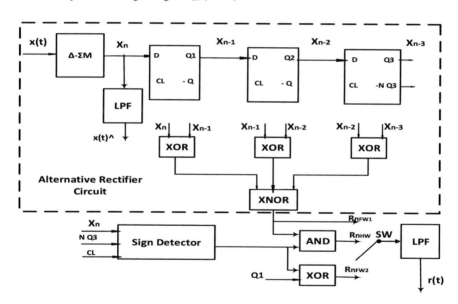

Fig. 3.17 Alternative version of the rectifier circuit

The proposed rectifier in Fig. 3.17 can perform as a half-wave rectifier when the switch of a filter is in position RnHW. The rectified waveforms are generated for the normalized input level of sinusoid $X = 0.01$. From Fig. 3.20, we see again a correct operation of a full-wave rectifier (no sign detector), and the half-wave rectifier (the sign detector included).

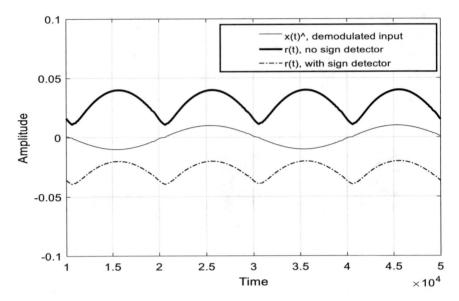

Fig. 3.18 Rectified waveforms of a sinusoid for normalized input level, $X = 0.01$

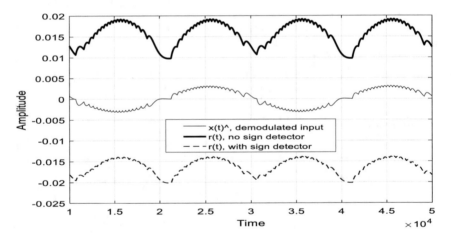

Fig. 3.19 Rectified waveforms of sinusoid for normalized input level, $X = 0.003$

3.7   Summary

Recent investigation of the behavior of three different rectifying encoders indicates that the RE with the sign detector circuit can be used for spike suppression in rectified first-order Δ-Σ modulated signals [14, 16]. Fujisaka et al. have shown that rectification of the first-order multi-level Δ-Σ is possible as is a second-order Δ-Σ [11]. The current investigation has shown that the RE with the sign detector circuit can be successfully used for complete suppression of spikes in the third-order Δ-Σ full-

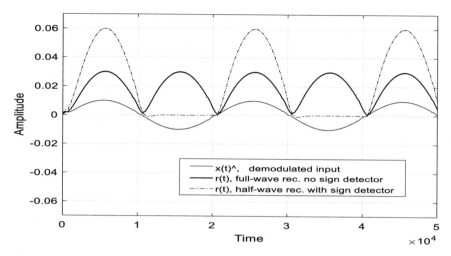

Fig. 3.20 Half- and full-wave rectified waveforms for input level $X = 0.01$

wave rectifying system. Waveforms and transitional behavior of a sign detector circuit have been analyzed. Adding two logic gates (XNOR and NAND) to the existing RE-sign detector scheme enables the proposed circuit to perform half-wave rectification of the third-order Δ-Σ modulated bit-stream. A proposed alternative gives promising results without use of a sign detector. The same circuit can operate as a squarer of a Δ-Σ bit-stream, when demodulating LPF has a lower cut-off frequency. The main contribution of this chapter is in presenting three novel circuits for rectification/squaring operations on Δ-Σ modulated bit-stream. Use of a full-wave rectifier is also not application limited. Currently, this type of circuit is used in instrumentation, communication receivers, and median filters [10]. However, the half-wave rectifier has more limited use. Potentially, it can be used in power supply equipment.

References

1. Kouvaras, N. (1978). Operations on delta-modulated signals and their application in the realization of digital filters. *Radio and Electronic Engineer, 48*(9), 431–438.
2. O'Leary, P., & Maloberty, F. (1990). Bit stream adder for oversampling coded data. *Electronics Letters, 26*(20), 1708–1709.
3. Liang, L., Wang, Z. G., Meng, Q. Q., & Guo, X. D. (2010). Design of high-speed high SNR bit-stream adder based on $\Sigma\Delta$ modulation. *Electronics Letters, 46*(11), 752–753.
4. Johns, D., & Lewis, D. (1993). Design and analysis of delta-sigma based IIR filters. *IEEE Transactions on Circuits and Systems II: Analog and Digital Signal Processing, 40*(4), 233–240.
5. de Fonte Dias, V. (1995). Signal processing in the sigma-delta domain. *Microelectronics Journal, 26*, 543–562.

6. Hidaka, Y., Fujisaka, H., Sakamoto, M., & Morisue, M. (2002). Piecewise linear operations on sigma-delta modulated signals. In *9th International Conference on Electronics, Circuits and Systems*, vol. 3, pp. 983–986.
7. Matsuyama, K., Fujisaka, H., & Kamino, T. (2005). Arithmetic and piecewise linear circuits for sigma-delta domain multi-level signal processing. In *2005 International Symposium on Nonlinear Theory and its Applications* (NOLTA 2005), October 2005, pp. 58–61.
8. Fujisaka, H. (2010). Single-electron circuit for sigma-delta domain signal processing, cutting edge nanotechnology. *INTECH*. www.intechopen.com, pp. 347–372.
9. Fujisaka, H., Kamio, T., Ahn, C.-J., Sakamoto, M., & Haeiwa, K. (2012). Sorter- based arithmetic circuits for sigma-delta domain signal processing - part I: Addition, approximate transcendental functions, and log-domain operations. *IEEE Transactions on Circuits and Systems I: Regular Papers, 59*(9), 1952–1965.
10. Fujisaka, H., Sakamoto, M., Ahn, C.-J., Kamio, T., & Haeiwa, K. (2012). Sorter-based arithmetic circuits for sigma-delta domain signal processing - part ii: Multiplication and algebraic functions. *IEEE Transactions on Circuits and Systems I: Regular Papers, 59*(9), 1966–1979.
11. Fujisaka, H., Kamio, T., Ahn, C.-J., & Haeiwa, K. (2013). Sequence characteristics of multi-level and second-order sigma-delta modulated signals. *IEICE Nonlinear Theory and Its Applications, 4*, 313–339.
12. Freedman, M., & Zrilic, D. G. (1990). Nonlinear arithmetic operations on the delta sigma pulse stream. *Signal Processing, 21*(1), 25–35.
13. Hein, S., & Zakho, A. (1993). *Sigma delta modulators, nonlinear decoding algorithms and stability analysis*. Boston: Kluwer. ISBN 978-1-4615-3138-8.
14. Zrilic, D. G., Petrovic, G., & Tang, W. (2017). Novel solution of a delta-sigma-based rectifying encoder. *IEEE Transactions on Circuits and Systems II: Express Briefs, 64*(10), 1242–1246.
15. Zrilic, D. G. (2012). US Patent, Appl. No. 13/694,560, Dec. 12, 2012.
16. Zrilic, D. G. *Method and apparatus for full-wave rectification of delta- sigma modulated signals*. US Patent No.: 9, 525, 430 B1.
17. Zrilic, D. G. (2015). US Patent, Appl. No. 14/999,971, Dec. 03, 2015.
18. Chao, K. C. H., Nadeem, S., Lee, W. L., & Sodini, C. G. (1990). A higher order topology for interpolative modulators for oversampling A/D converters. *IEEE Transaction on Circuits and Systems, CAS-37*, 308–318.
19. de la Rosa, J. M. (2011). Sigma-delta modulators: Tutorial overview, design guide, and state of the art survey. *IEEE Transactions on Circuits and Systems I: Regular Papers, 58*(1), 1–2.
20. Gray, R. (1987). Oversampled sigma-delta modulation. *IEEE Transactions on Communications, 35*(5), 481–489.
21. Bourdopoulos, G. I., Pnevmatikakis, A., Anastassopoulos, V., & Deliyannis, T. L. (2006). *Delta-Sigma modulators: Modeling, design and applications*. London: Imperial College Press. ISBN 1-86094-369-1.

Chapter 4
Multiplication of Two Δ-Σ Bit-Streams

4.1 Introduction

In spite of the fast progress of semiconductor technology, there are still a number of open problems with n-bit digital signal processing (DSP). For example, flash analog-to-digital converters are bulky and power hungry. Digital multiplying circuits are bulky and power hungry as well. This problem becomes more acute when we deal with applications which require a 20-bit or more resolution. Length of the code word can be efficiently reduced by using a differential pulse-code modulation (DPCM). Δ-ΣM is a one-bit DPCM system and employs a trade-off between a number of amplitude quantization levels and sampling frequency.

As explained earlier, Δ-ΣM analog-to-digital converters are characterized by one-bit quantization and a very high sampling rate [1]. To process a Δ-ΣM pulse stream with ordinary DSP hardware decimation is needed first [2]. Arithmetic operations are then performed with DSP hardware. Usually these converters are integrated with complex decimation filters on the same chip. There are many applications in control, robotics, instrumentation, industrial processes, etc., where direct processing of a Δ-ΣM pulse stream is preferred. One of the most important advantages of a Δ-ΣM serial output is the possibility of manipulating the serial output in the digital domain.

Neither the literature nor practice fully recognizes yet the possibility of direct arithmetic operations on a Δ-ΣM pulse stream. Pioneering work in this area was done by Kouvaras [3]. Multipliers of delta-sigma sequences were proposed in reference [4] by Lagoyannis and Pekmetzi. Three such circuits of multipliers are proposed with varying circuit complexity and performance. The operation of a proposed multiplier is based on the use of a delta adder and rectifying encoder described in Chaps. 2 and 3. Consequently, we will briefly describe the operation of two newly proposed multiplier circuits.

© The Editor(s) (if applicable) and The Author(s), under exclusive license to
Springer Nature Switzerland AG 2020
D. Zrilic, *Functional Processing of Delta-Sigma Bit-Stream*,
https://doi.org/10.1007/978-3-030-47648-9_4

4.2   Δ-Σ Adder

Analytical evaluation of formulas defined by Kouvaras was presented in Chap. 2 of this monograph. The resulting signal of the sum of an adder is defined by Kouvaras as:

$$S_n = 1/2[X_n + Y_n] - 1/2[C_{n-1} - C_n] \tag{4.1}$$

$$C_n = X_n Y_n C_{n-1}, \tag{4.2}$$

where X_n, Y_n, C_n, C_{n-1}, and S_n take the values of +1 and −1.

These relations lead Kouvaras to the synthesis of a Δ-Σ adder with interchanged roles of the sum and carry output of a binary full adder. When the binary sequence S_n is demodulated, one-half of the sum of signals $x(t)$ and $y(t)$ is obtained

$$\hat{s}(t) = 2^{-1}[(x(t) + y(t)] - 2^{-1}[e_1(t) + e_2(t)] + \phi(t), \tag{4.3}$$

where $2^{-1}[e_1(t) + e_2(t)]$ is the half-sum of the error of the two Δ-Σ systems and can be considered as the equivalent error of the Δ-Σ system, the input of which is the analog signal $2^{-1}[x(t) + y(t)]$. The value of $\varphi(t)$, because of introduction of a full adder, is $|\varphi(t)| \leq \delta$, where δ is the step size of a delta modulator.

4.3   Rectifying Encoder

A simple rectifying encoder is shown in Fig. 3.1 of Chap. 3. This circuit can perform a squaring operation of delta-modulated bit-stream when demodulating low-pass filter has cut-off frequency nearly 2fin. Figure 3.10 illustrates the case of a squaring operation of sinusoidal signal of the system shown in Fig. 3.9. Applying a sum or difference of two analog signals to the delta-sigma modulator and then squaring them using RE will produce sum $S_{n1} = (X_n)^2 + 2X_n Y_n + (Y_n)^2$ or $S_{n2} = (X_n)^2 - 2X_n Y_n + (Y_n)^2$ bit-streams. Subtracting S_{n2} from S_{n1}, the resulting signal is $S_n = 4X_n Y_n$. Demodulating the sum signal S_n by a low-pass filter of proper cut-off frequency analog signal $s(t)$ is obtained. Thus, the operation of the multiplier is based on the delta adder and the rectifying encoder. In the following section, two approaches of multiplication are presented. The first approach is based on the analog summation and subtraction of input signals to the delta modulators, while the second approach is based on a fully digital solution.

4.4   Multiplication of Two Δ-Σ Bit-Streams

A. *Approach #1*

A block diagram of a proposed multiplication circuit of two analog signals $x(t)$ and $y(t)$ by means of Δ-ΣM is shown in Fig. 4.1 [5]. As can be seen, the sum and difference of analog signals $x(t)$ and $y(t)$ are fed into two synchronous Δ-Σ modulators. Digital output bit-streams $X_n + Y_n$, and $X_n - Y_n$, of delta modulators are delayed by one clock pulse and then added in XOR gates. The D flip-flop with an XOR gate is a differentiator of the delta-modulated pulse stream, producing the square (quadratic) signal at its output. Thus, the output of XOR$_1$ gate is $(X_n + Y_n)^2$ and the output of XOR$_2$ gate is $(X_n - Y_n)^2$. The outputs of the XOR gates are then fed into a delta adder (with $X_n - Y_n$ input inverted) to obtain the sum

$$S_n = \left(X_n + Y_n \right)^2 - \left(X_n - Y_n \right)^2 = 4X_n Y_n. \tag{4.4}$$

However, attenuation of a delta adder must be considered as well.

After demodulation (averaging) of sequence S_n, the product of $x(t)*y(t)$ is obtained. Figure 4.2 shows simulation results of multiplication of near signals, $x(t) = \sin(t)$ and $y(t) = e^{-t}$, at sampling time $T = 0.00002$ s, and the length of the averaging filter is $N = 1000$ samples. Existing delay is caused by the filter (averaging).

In Fig. 4.3, the results of simulations are shown for the case when $x(t) = y(t) = e^{-2t}$.

B. *Approach #2*

To avoid analog addition of input signals and eventual overload of delta modulators we propose a digital solution with a delta adder as shown in Fig. 4.4 [6]. Analog input signals $x(t)$ and $y(t)$ are converted by synchronous delta-sigma modulators to produce bit-streams Xn and Yn. These signals are added and subtracted to produce bit-streams $Xn + Yn$ and $Xn - Yn$, respectively.

After the squaring operation, these signals are subtracted in a ΔA3 to produce signal Sn. After demodulation signal $p(t)$ is obtained. We must take into account attenuation of delta adders and low-pass filter as well. Thus, we can write that $p(t) = 4kx(t)*y(t)$, where k is attenuation constant. In Fig. 4.5, theoretical and simulation waveforms of multiplication of two sinusoidal signals (*fx* = 9 Hz and

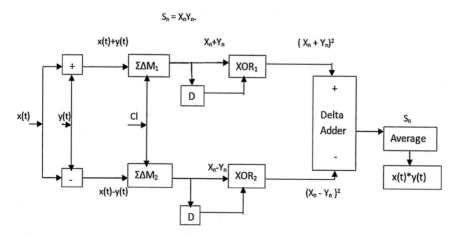

Fig. 4.1 A proposed multiplication circuit, Approach #1 [5]

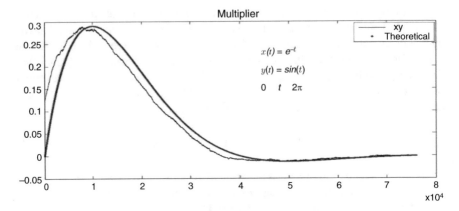

Fig. 4.2 Theoretical and simulation result of multiplication

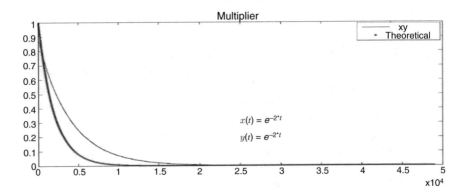

Fig. 4.3 The case of multiplication when $x(t) = y(t)$

$fy = 18$ Hz), with normalized amplitudes of 0.8 and 0.6. We see a close agreement between theoretical and simulation waveforms.

The circuit presented in Fig. 4.4 can serve as a squarer of an input signal as well. When $x(t) = y(t)$, then output of Δ-ΣM2 generates an idle sequence, which after low-pass filtering produces a zero-output signal. An example of a squaring operation is shown in Fig. 4.6.

4.5   Summary

We have demonstrated how multiplication of two Δ-ΣM bit-streams can be implemented using inexpensive, off-the-shelf components such as D flip-flops, logic gates, and adders (or using FPGA). It is suitable for an IC design as well. The proposed circuits differ from circuits proposed in reference [4] and can be extended

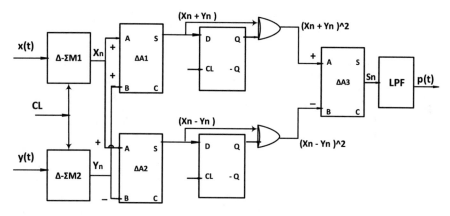

Fig. 4.4 A proposed multiplication circuit, Approach #2 [6]

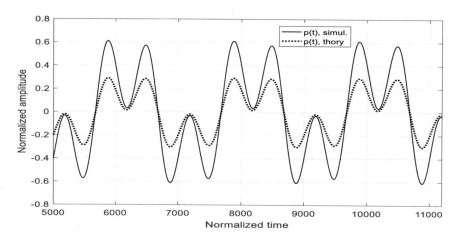

Fig. 4.5 Multiplication of two sinusoidal signals

easily for multiplication of three or more delta-modulated sequences. The main building blocks of this circuit are the delta adder and the rectifying encoder. We demonstrate the possibility of direct multiplication of Δ-ΣM sequences without use of bulky decimation filters. There are a number of remaining research tasks such as determining the optimal levels of input signals, the optimal length of an averaging filter, and the optimal oversampling ratio. In general, error analysis of the proposed novel multiplication systems should be performed. The main contribution of this chapter is in presenting two novel multiplying circuits whose operation is based on linear/nonlinear operations on Δ-Σ modulated bit-streams.

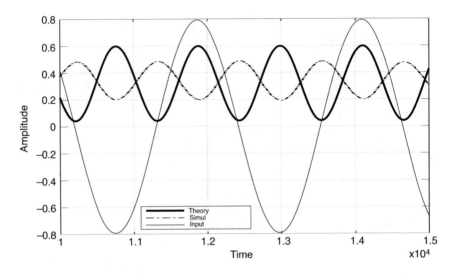

Fig. 4.6 Squaring of sinusoidal input

References

1. Candy, J., & Temes, G. (Eds.). (1992). *Oversampling delta-sigma data converters, theory, design and simulations*. New York: IEEE Press. ISBN 0-87942-285-8.
2. Norsworthy, S., Schreier, R., & Temes, G. (Eds.). (1997). *Delta-sigma data converter*. New York: IEEE Press. ISBN 7803-1045.
3. Kouvaras, N. (1978). Operations on delta-modulated signals and their applications in the realization of digital filters. *The Radio and Electrical Engineer, 48*(9), 431–438.
4. Lagoyannis, D., & Pekmetzi, K. (1981). Multipliers of delta-sigma sequences. *The Radio and Electronic Engineer, 51*(6), 281–286.
5. Zrilic, Đ. G. (2010, September 19–23). Circuit for multiplication of two sigma-delta modulated pulse streams. In *The CD of the 9th World Automation Congress*, Kobe, Japan.
6. Zrilic, Đ. G. *US PTO*, Pub. No: 2014/0159929, Appl. No: 13/694,560.

Chapter 5
Digital Architecture for Delta-Sigma RMS-to-DC Converter

5.1 Introduction

Δ-Σ modulation is a well-established analog-to-digital conversion (ADC) process. It is a low-power consuming, high-resolution, one-bit conversion process, and it is suitable for VLSI design. It has applications in low-frequency ADC processes such as biomedical applications, environmental monitoring, seismic, instrumentation, etc. It also has applications in audio and radio frequencies. The RMS-to-DC Δ-ΣM circuit can be used for automatic-gain control (AGC) of the amplifier to maintain a constant output level of varying waveform. The RMS-to-DC Δ-ΣM instrument can be used as a low-cost true RMS digital panel meter for direct measurement of power consumption in different household appliances such as stoves, TV sets, refrigerators, etc. It can be implemented as an AC line-powered version. The RMS-to-DC Δ-ΣM circuit can be used as a portable, high-impedance input RMS panel meter and dB meter for a modem line monitor. The RMS-to-DC Δ-ΣM can be used in micro-grid power lines metering and mobile communication radio frequency level monitoring. In addition, the RMS-to-DC Δ-ΣM circuit has applications in data acquisition systems for detection of a signal level, or testing and grading components such as transistors, op amplifiers, and many others.

There are a number of published references dealing with integrated RMS-to-DC converters. Most of these publications deal with analog processing methods using trans-linear properties of bipolar circuitry. These converters are difficult to implement in standard CMOS processes, however. In digital processing methods, which use high-speed n-bit ADC, the RMS value of an input signal is calculated by ordinary DSP calculation methods. The problem is that the digital computation unit occupies a large chip area and thus does not lead to a low-cost solution. To cope with analog component limitations, the use of Δ-Σ modulation is proposed in reference [1], Fig. 5.1. A similar approach is used by Linear Technology Inc. [2]. In both approaches, an analog signal is Δ-Σ modulated and then multiplied (in a multiplying

© The Editor(s) (if applicable) and The Author(s), under exclusive license to
Springer Nature Switzerland AG 2020
D. Zrilic, *Functional Processing of Delta-Sigma Bit-Stream*,
https://doi.org/10.1007/978-3-030-47648-9_5

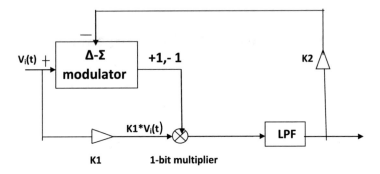

Fig. 5.1 The Δ-Σ true RMS converter [1, 2]

DAC) by a polarity signal at the output of a Δ-Σ modulator. This mixed processed signal is then filtered and fed back to the negative input of the Δ-Σ modulator as a reference signal. When the reference input signal of a Δ-Σ modulator is taken into account, then the output of the modulator can be interpreted as a ratio-metric function plus quantization noise (US Patent Number: 6,587,061 B2) [2]. The mixed mode analog/digital multiplier, which performs the squaring operation of the analog signal by means of Δ-ΣM, is known as a polarity switch and is proposed in reference (US Patent Number: 6,285,306 B1) [3].

Analysis of a Δ-Σ multiplier–divider, using a second-order Δ-Σ modulator, is presented in reference [1]. We have performed simulations for both the first- and second-order Δ-Σ RMS-to-DC converters. For completeness, we will briefly introduce a principle of division by a Δ-Σ modulator of the first order.

5.2   The First-Order Δ-Σ Divider

In our analysis of a linearized system, we assume a busy input signal and quantization error is considered to be a white Gaussian noise. A linear Δ-Σ model in Fig. 5.2 can serve as a divider of two signals under the condition that a denominator must be positive and slow changing. First let us derive a basic equation for a quantizer output d_n. Applying the principle of superposition, we can write

$$d_{n1} = u_n - u_{n-1}, \text{for} V_n = 0 \tag{5.1}$$

$$d_{n2} = V_{n-1}, \text{for} u_n = 0 \tag{5.2}$$

$$d_n = d_{n1} + d_{n2} = u_n - u_{n-1} + V_{n-1}. \tag{5.3}$$

We see that the quantizer output dn consists of quantization error $q = u_n - u_{n-1}$ and input signal V_{n-1}.

Similarly, the output signal of a divider ($W_n > 0$) can be written as

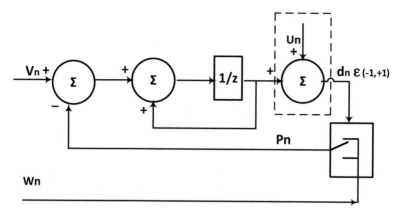

Fig. 5.2 A linear model of the first-order Δ-Σ divider

$$d_n = V_{n-1}/W_n + U_n - \left(W_{n-1}/W_n\right)U_{n-1}. \tag{5.4}$$

Because $W_n = W_{n-1} = $ constant $= W$, we have

$$d_n = V_{n-1}/W + U_n - U_{n-1}. \tag{5.5}$$

Taking a Z-transform of Eq. (5.5) we have

$$D(z) = \left(V(z)z^{-1}\right)/W + U(z)\left(1 - z^{-1}\right). \tag{5.6}$$

For a high sampling ratio, $fs/fN \gg 1$, the second term in Eq. (5.6) can be neglected and the transfer function of the system is

$$D(z) = \left(V(z)z^{-1}\right)/W. \tag{5.7}$$

Applying this simplified analysis to the proposed Δ-ΣM RMS-to-DC converter in Fig. 5.3, one can obtain an expression for the effective value of the input signal. Following notation in Fig. 5.3 we can write:

$$V_{dc} = \text{ave}\left(Y_n\right) = \text{ave}\left|d_n\right| \tag{5.8}$$

$$d_n = V_n/W_n \tag{5.9}$$

$$V_{dc} = \text{ave}\left|V_n/W_n\right|. \tag{5.10}$$

Because W_n is in a steady state (nearly a DC value) and proportional to the amplitude V_{dc}, we can write

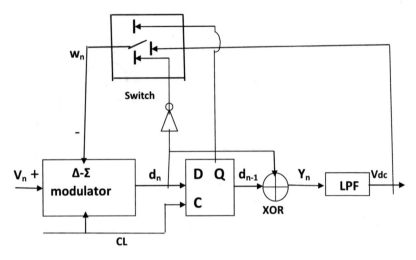

Fig. 5.3 The proposed Δ-ΣM RMS-to-DC converter

$$W_n = V_{dc} \tag{5.11}$$

$$V_{dc} = \mathrm{ave}\left(V_n/V_{dc}\right)^2 = \mathrm{ave}\left(V_n\right)^2 / V_{dc}. \tag{5.12}$$

Equating (5.10) with (5.12) we have

$$V_{dc} = \text{square root}\left(\text{average}\left(V_n\right)^2\right). \tag{5.13}$$

Unlike prior RMS-to-DC converters [1, 2], the proposed converter performs a direct nonlinear operation on a Δ-Σ bit-stream using a rectifying encoder [4]. In addition, it does not use analog multiplication and amplification of a feedback signal.

5.3 Simulation Results

Verification of a proposed RMS system is performed by simulation for both first- and second-order Δ-Σ modulation systems. In Fig. 5.4, the V_{rms} value is shown for a sinusoidal input signal, normalized amplitude $V_{in} = 0.01$, when a first-order Δ-Σ modulator is employed as an ADC.

Under the same simulation condition Fig. 5.5 is generated using a second-order Δ-Σ modulator as an ADC.

In Fig. 5.6, a transfer characteristic of the proposed system in Fig. 5.3 is shown. When a modulator is properly oversampled ($fs \gg fN$), we see a close agreement between theoretical and simulation transfer characteristics.

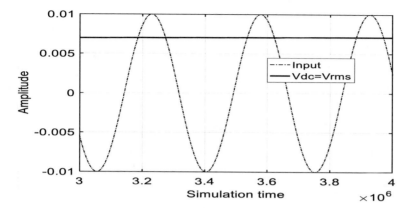

Fig. 5.4 Simulation result for the first-order Δ-Σ RMS

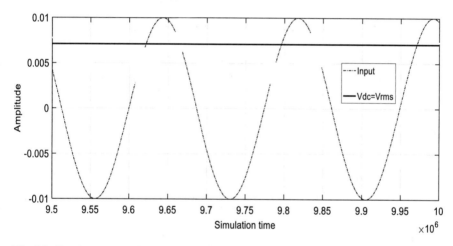

Fig. 5.5 Simulation results for the second-order Δ-Σ RMS

5.4   Summary

Root mean square (RMS) is a fundamental measurement of the magnitude of an alternate current (AC) signal. Its definition can be both mathematical and practical. The mathematical formula involves squaring the signal, taking the average, and obtaining the square root. The averaging time must be sufficiently long to allow filtering at the lowest frequencies of the operation desired. Practical definition: the RMS value assigned to an AC is the amount of direct current (DC) required to produce an equivalent amount of heat in the same load [5]. Implementation of a newly proposed Δ-Σ RMS-to-DC system is based on a direct rectifying operation of a Δ-Σ bit-stream and inherent ability of Δ-Σ to perform a division operation and an implicit calculation of the RMS value of an input signal. The fundamental building block is

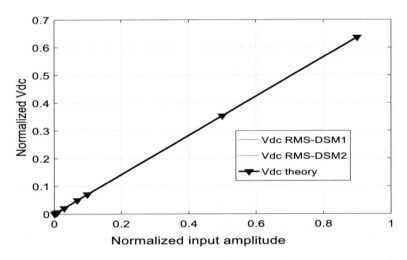

Fig. 5.6 Transfer characteristic of the proposed Δ-Σ RMS-to-DC converter

the rectifying encoder of the Δ-Σ bit-stream. The proposed Δ-Σ RMS circuit is suitable for a VLSI design, with a capacitor as an external component.

References

1. Wey, W.-S., & Huang, Y.-C. (2000). A CMOS delta-sigma true RMS converter. *IEEE Journal of Solid-State Circuits, 35*(2), 248–257.
2. Petrofsky, J., & Brisebois, G. (2002). Δ-Σ breakthrough: LTC 1966 true RMS-to-Dc converter uses no diodes, hearers or logarithms. *Linear Technology Magazine, XII*(1), 1–16.
3. Zrilic, D. G. (2016). US patent, Appl. No: 14/999,288, Filed: April 21, 2016.
4. Zrilic, D. G., Petrovic, G., & Tang, W. (2017). Novel solution of a delta-sigma-based rectifying encoder. *IEEE Transactions on Circuits and Systems II: Express Briefs, 64*(10), 1242–1246.
5. Kitchin, C. (1986). *L counts, RMS-to-DC conversion application guide* (2nd ed.). Norwood, MA: Analog Devices Inc.

Chapter 6
Companding Circuits and Systems Based on Δ-Σ Modulation

6.1 Introduction

There are several publications and patents dealing with companding of analog signals. Most of these publications deal with analog implementation of companding circuits, which are used in telephone transmission systems. In the past six decades, pulse-code modulation (PCM) was used as an analog-to-digital converter (ADC) in digital telephone systems. The PCM encoding schemes are recommended by the International Telecommunication Union (ITU) and are the international PCM companding standards. European countries practice logarithmic **A**-Law, while logarithmic **μ**-law companding technique is used in North America and Japan. A brief introduction about the need for companding in PCM-based digital telephone systems can be found in references [1–3]. In addition, there are a number of communication books describing in detail the operation of PCM technique [4]. Existing compander systems are composed of complex analog circuits that provide good sound quality. However, analog circuits are, by their nature, subject to variable performance, and use of advanced techniques is required to keep performance levels constant. There are compander integrated circuit (IC) chips on the market [5, 6]. Their implementation is analog in nature.

A novel square-law compander architecture, based on Δ-Σ modulation technique for telecom application, is discussed in reference [7]. The authors of the paper claim that the proposed compressor and expander circuits reduce the number of off-chip components. However, from Figs. 2 and 3 [7], we see that both compressor and expander consist of several analog components.

In reference [8] a novel technique is proposed, based on the use of a Δ-Σ modulator and operational transconductance amplifier (OTA). Even though this implementation of compressor and expander circuit is simple (Fig. 4 in [8]), OTA is an analog circuit. In addition, an adaptive delta modulator is implemented with 6 OTAs, 2 diodes, 2 resistors, and 3 capacitors (Fig. 2 in [8]). As with a standard operational

© The Editor(s) (if applicable) and The Author(s), under exclusive license to
Springer Nature Switzerland AG 2020
D. Zrilic, *Functional Processing of Delta-Sigma Bit-Stream*,
https://doi.org/10.1007/978-3-030-47648-9_6

amplifier, practical OTAs have some non-ideal characteristics, such as input stage nonlinearity at higher differential input voltages, temperature sensitivity of trans-conductance, variation of input and output impedance with control current, and bias voltages as well. The Δ-Σ based CMOS compander circuit is proposed in [9]. The proposed configuration is a suitable alternative to the conventional compander structures. However, the proposed compander's envelope detector circuit is complex (Fig. 6 in [9]). It is implemented with switch capacitor (SC) technology and the entire compander operates at 500 kHz. Almost all digital implementation of a square-law compander is proposed in [10]. In addition to a RC low-pass filter, there are two comparators. Victor da Fonte Dias [11] proposed a number of circuits for signal processing in the delta-sigma domain. In all the proposed circuits (compressor, expander, and compander) the rectification circuit is implemented in an analog domain. LPF and rectifiers are implemented as external components of a Δ-Σ modulator. Having in mind current development in the area of compression and expansion of analog signals, we are proposing a novel technique based on direct processing of a Δ-Σ bit-stream. Depending on the application, proposed circuits can be integrated into an IC chip.

6.2 Background

Telephone pulse-code modulation (PCM) systems have been in use since 1960. The process of PCM analog-to-digital conversion (ADC) is comprised of three phases: sampling, quantization, and coding. A process of quantization can be linear or non-linear. Dynamic ($D = 20 \log(Amx/Amin)$) of human speech is large, and it is proven that quitter phonemes (utterance and sounds) are more probable to occur and carry more information than louder ones. Thus, it is natural to amplify smaller amplitude and attenuate larger in order to reduce the dynamic range, i.e., a number of bits per sample, transmitted over a 4 kHz telephone network. Use of a linear PCM encoding system is not acceptable, and use of companding technique, which reduces the number of bits per sample, makes a good fit. The word "compander" is a portmanteau for a "compression–expanding" technique. It compresses the signals before transmitting on a band-limited channel and expands them when received. During compression smaller amplitudes are amplified and larger amplitudes are attenuated. Conversely, at expander, a digital signal is converted back to an analog signal, and low-level amplitudes are amplified less when compared to higher ones. In short, altering the dynamic range of the signal is the key role of a compressor. Even though both transfer functions are nonlinear, the overall transfer function "compressor–expander" system must be linear. In addition to the PCM-based telephone systems, compression/expansion technique finds application in sound recording, TV, radio, and public address systems. For readers interested in this topic we quote some relevant references, but there is a wealth of information available on the Internet as well.

In the following sections we will introduce novel circuits of compressor, expander, compander, and a digital system for post-processing of the audio signal. In addition, simulation results are presented.

6.3   Δ-Σ Compressor

The operation of a proposed compressor circuit, shown in Fig. 6.1, is based on direct nonlinear processing of a Δ-Σ bit-stream. Δ-Σ pulse stream Xn is rectified using rectifying encoder (RE), which produces bit-stream Yn, and then low-pass filtered to obtain a nearly DC signal $y(t)$. Signal $y(t)$ is inverted to obtain signal-$y(t)$, which is multiplied by polarity signal D to obtain signal $a(t)$, which is fed into negative input of a second-order Δ-Σ modulator. A feedback signal $a(t)$ is shown in Fig. 6.2.

We can see an embedded polar signal D into an envelope signal $y(t)$. Figure 6.3 shows both an input signal $x(t) = e^{-t}\sin\omega t$ and its compressed version $c(t)$.

We see that the larger input amplitudes are slightly attenuated and distorted, while the lower input amplitudes are amplified. Thus, the dynamic range of the compressed signal is reduced, and consequently a number of bits per sample are reduced.

In Fig. 6.4, a quadratic function of compressor, $c(t)$ = square root of $(x(t))$ is shown. There is close agreement between theoretical and simulation results.

6.4   Δ-Σ Expander

The expander circuit proposed in [9, 11] is composed of a delta-sigma modulator, a rectifier circuit, a low-pass filter, and a multiplier. The input signal is delta modulated and at the same time, in parallel, is rectified with analog components. This

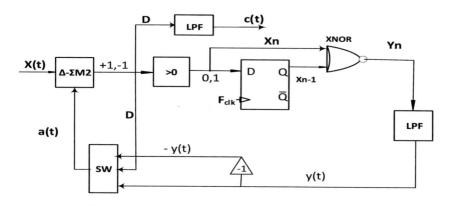

Fig. 6.1 A proposed Δ-Σ compressor circuit [12]

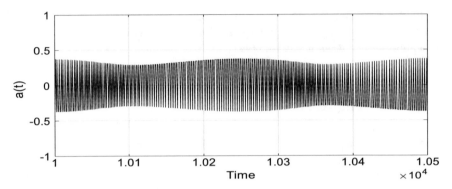

Fig. 6.2 A feedback signal $a(t)$

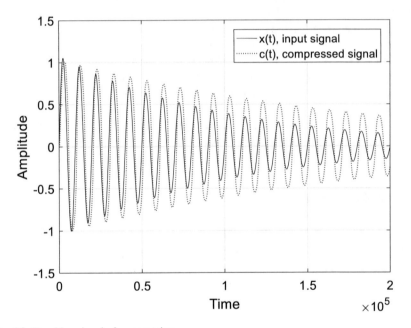

Fig. 6.3 Resulting signal of compression

externally rectified input analog signal is multiplied by means of a switch with a delta-sigma modulated bit-stream. The output of a switch is low-pass filtered. Figure 6.5 shows the proposed expander circuit, where rectification of an analog input signal is performed on a delta-sigma modulated bit-stream. Again, we see that there is no significant difference between compressor and expander circuits. The only difference is that the expander circuit is an open-loop system, while the compressor circuit is a closed-loop system, where signal $a(t)$ is fed into a Δ-Σ modulator. Depending on the application, the expander circuit can operate as a stand-alone or in connection with compressor. If operated as a stand-alone system, then special attention must be paid to the level of the analog input signal.

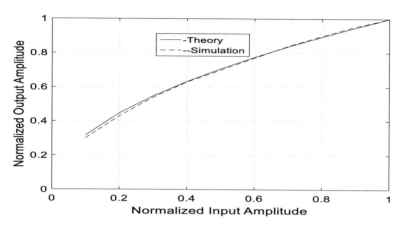

Fig. 6.4 Transfer function of compressor

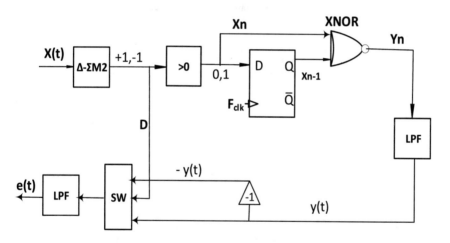

Fig. 6.5 A proposed Δ-Σ expander circuit [12]

When the analog compressed signal $c(t)$, Fig. 6.1, is delivered to the input of a Δ-Σ modulator of an expander in Fig. 6.5, the identical signal is obtained ($x(t)$ at the input of compressor and $e(t)$ at the output of expander) (Fig. 6.6).

A quadratic transfer theoretical and simulation functions of an expander are shown in Fig. 6.7. We can see close agreement between both functions.

6.5 Δ-Σ Compander

A novel delta-sigma based square-law compander circuit was presented in reference [7]. Both compressor and expander circuit have a number of analog blocks such as a polarity detector, rectifier with LPF, level detector, and two clock generators,

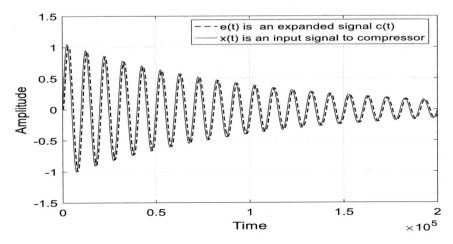

Fig. 6.6 Expanded signal $e(t)$ and original input signal to compressor $x(t)$

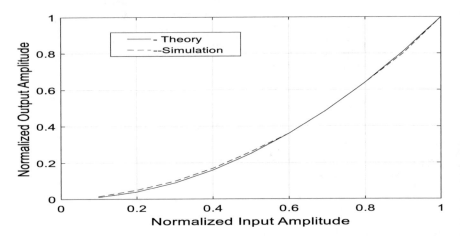

Fig. 6.7 Theoretical and simulated transfer function of expander

where a slow clock generator is regulated with a polarity detector. To overcome some of these problems, we propose a novel square-law compander. Joint operation of a compressor and an expander is frequently used in practice, rather than a stand-alone operation of compressor or expander. For example, a linear PCM encoder is preceded by a compressor, which reduces a dynamic of input signal, i.e., a number of bits per PCM sample. This PCM signal is transmitted and at the receiver decoded and expanded to obtain the original signal. Figure 6.8 shows back-to-back connection of a compressor and an expander circuit. In this case a digital compressed signal D is directly connected to the RE of an expander, and there is no need for a Δ-Σ modulator at the receiving side.

  Figure 6.9 shows three relevant waveforms of a back-to-back connection. To show the validity of operation, compressed digital signal D is demodulated at the transmitting side (signal $c(t)$). We see that input signal $x(t)$ is identical to the expanded signal $e(t)$.

Even though both transfer functions are highly nonlinear, overall the transfer function of a compander circuit is linear. In Fig. 6.10, a transfer function of compander circuit is shown.

There are applications where post-processing of a compressed signal is desired. For instance, Δ-Σ modulation is frequently used because of a high resolution in the audio recording industry where digital information can be easily stored and manipulated to achieve certain sound effects. In Fig. 6.11, we propose a novel system for audio recording.

As can be seen, analog sum A is compressed and then manipulated (delayed) to achieve the desired studio effect. Digital signal C is expanded and fed back to the

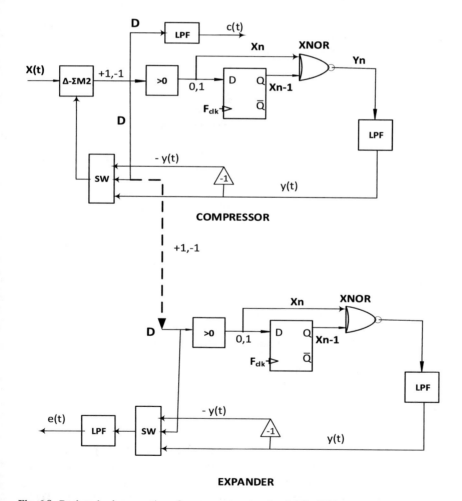

Fig. 6.8 Back-to-back connection of compressor–expander circuits [12]

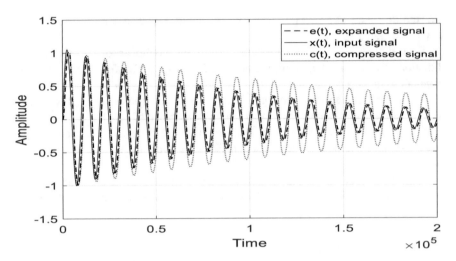

Fig. 6.9 Relevant waveforms of the compander circuit

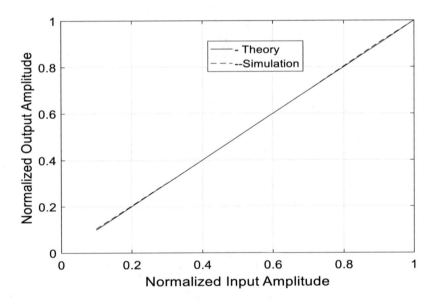

Fig. 6.10 Compander's transfer function

input (signal $y(t)$). At the same time an expanded signal $y(t)$ is added to the original signal $x(t)$. Combined signal E is amplified/attenuated to obtain the desired output signal.

Traditional solid-state bucket brigade devices (BBD) are used to provide a signal delay [6]. The main drawbacks of these devices are degradation of the capacitor charge and appearance of spikes in the wave that correspond to the clock frequency. In Fig. 6.12, we propose a simple programmable delay line which consists of D flip-flops only.

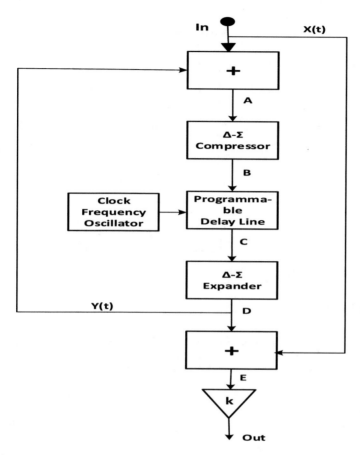

Fig. 6.11   Δ-Σ audio recording companding system [12]

6.6   **Summary**

Four novel circuits are presented for compression, expansion, companding, and post-processing of a compressed delta-sigma bit-stream. Compression and expansion are based on the use of a second-order (or higher-order) delta-sigma modulator and a nonlinear operation on a delta-sigma modulated bit-stream. Depending on the application, the proposed circuits can operate as a stand-alone integrated circuit, or as compander, as proposed. An inherited low-pass filter can be digital or analog; thus, the only external analog component to the IC chip could be capacitor C, when a low-frequency analog signal is compressed or expanded. The key advantage of the proposed circuits is their integrability in an IC circuit and other signal processing functions of a particular system. Depending on implementation technology there are additional open questions to be researched (answered) such as distortion analysis of compressor/expander over a wide input signal dynamic range, instability due to offset and noise, square-low versus linear gain operation, etc.

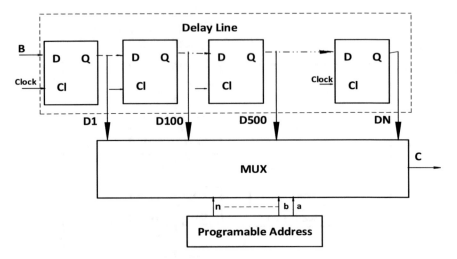

Fig. 6.12 A programmable delay line

References

1. Sneha, H. L. *Companding logarithmic laws, implementation, and consequences*. http://www.allaboutcircuits.com/technical-articles/companding
2. Sneha, H. L. *An introduction to companding: compressing speech for transmission across telephone systems*. http://www.allaboutcircuits.com/technical-atricles/an-introduction
3. Rudolph, B. *Understanding audio compressors and audio compression*. http://www.barryrudolph.com/mix/comp.html
4. Taub, H., & Schilling, D. L. (1986). *Principles of communications systems* (2nd ed.pp. 213–219). New Delhi: McGraw Hill.
5. Wiltshire, T. *Noise reduction with companders*. http://electricdruid.net/noise-reduction
6. Bucket Brigade Devices: MN3007 (/mn3007-bucket-brigade-devices). Download from Internet.
7. Takasuka, K. et al. (1990). A sigma-delta based square-law compandor. In *Proceedings CICC*, Boston, MA, pp. 12.7.1–12.7.4.
8. Roy, A., et al. (2010). Processing of communication signal using operational transconductance amplifier. *Journal of Telecommunications, 1*(1), 118–122.
9. Qiuting, H. (1992). Monolithic CMOS compandors based on Σ-Δ oversampling. In *Proceedings of IEEE ISCAS*, pp. 2649–2651.
10. Zrilic, D. G. (2005). *Circuits and systems based on delta modulation* (pp. 174–182). Berlin: Springer.
11. da Fonte Dias, V. (1996). Signal processing in the delta-sigma domain. *Microelectronics Journal, 26*(6), 543–562.
12. Zrilic, D. G. US patent Appl. No. 16/350, 315.

Chapter 7
A Δ-Σ Digital Stereo Multiplexing–Demultiplexing System

7.1 Introduction

Frequency modulation (FM) stereo broadcasting technique is well understood in literature and practice and is regulated by federal law (FCC). This stereo broadcast is not limited to FM; the FCC has authorized some form of stereo AM broadcast as well. There are a number of books and references describing the FM stereo technique. In Fig. 7.1 a generic block diagram of a stereo FM system is shown [1]. Similar block diagrams of a stereo FM system can be found in many other references and on the Internet. Thus, following general notation in this figure, we will briefly describe the operation of the system and point out its disadvantages. The idea of stereo is to provide a sound wave-front which replicates the depth and realism that an individual experiences when listening. As can be seen in Fig. 7.1, the signal broadcast consists of two channels: left (L) and right (R). This signal can be derived from a microphone, tape, or CD player, or another source. In the case of a broadcast of speech or music, the frequency bandwidth is from 30 Hz to 15 kHz. With FM stereophonic broadcasting, voice or music channels are frequency division multiplexed onto a single FM carrier. The L and R audio channels are combined in analog adder networks to produce the (L − R) and (L + R) audio channels. As one can see, the (L + R) signal is used to modulate the carrier just as a non-stereo signal does. The (L − R) signal is shifted by a balance modulator (subcarrier frequency is 38 kHz) to produce a double sideband (DSB) suppressed carrier (SC) signal spanning $38 - 15 = 23$ kHz and $38 + 15 = 53$ kHz. This process of stereo multiplexing is known as MPX. For demodulation purposes (synchronization), a 19 kHz pilot signal is also transmitted. All three signals are combined and delivered to an FM radio frequency transmitter. Most FM transmitters now use an integrated circuit (IC). Motorola MC1376 IC is a complete FM modulator. Unfortunately, it requires several external analog components to make it operate (including two inductor (coil) components). Similarly, the XR-1310 stereo demodulator has a significant

© The Editor(s) (if applicable) and The Author(s), under exclusive license to
Springer Nature Switzerland AG 2020
D. Zrilic, *Functional Processing of Delta-Sigma Bit-Stream*,
https://doi.org/10.1007/978-3-030-47648-9_7

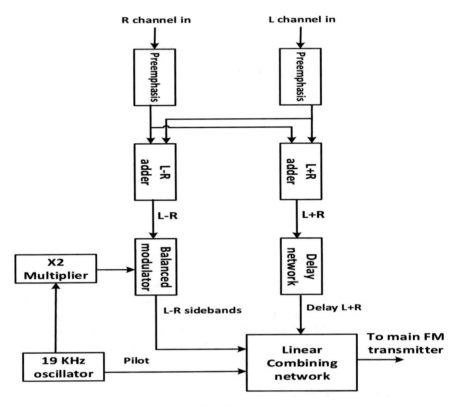

Fig. 7.1 General block diagram of an analog FM system

number of analog components. Thus, FM IC chips have to be buffered by a number of analog components, which it is not possible to integrate. There are several US Patents proposing different digital stereo methods, but they are not related to the digital stereo multiplexing–demultiplexing method we are proposing.

The proposed method of digital stereo multiplexing uses a high-resolution delta-sigma modulator (resolution ≥ 20 bits) as the analog-to-digital converter. A novel method of stereo multiplexing and creation of left and right digital channels is based on non-conventional digital signal processing using a delta adder. Operation of this system and simulation results will be shown in the next section.

7.2   Proposed Δ-Σ Modulation Stereo Multiplexing–Demultiplexing System

A stereo multiplexing–demultiplexing system is the main component of any stereo transmission system. Most existing stereo techniques are based on a frequency multiplexing technique. The proposed method, shown in Fig. 7.2, consists of a

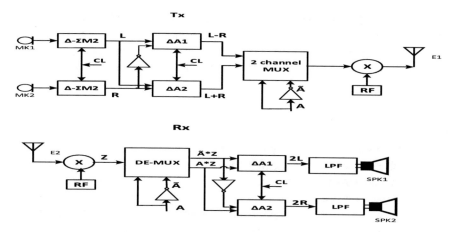

Fig. 7.2 A block diagram of Δ-Σ digital stereo multiplexing–demultiplexing system [2]

high-resolution Δ-Σ modulator whose resolution is greater than 20 bits [2]. In addition to the simple and inexpensive high-resolution Δ-Σ modulator implementation [3, 4], it is possible to perform arithmetic operations of addition, subtraction, and multiplication by a constant less than 1 on its bit-stream [5]. To implement addition/subtraction of the left (L) and right (R) channels a special circuit must be developed. Delta adder is an ordinary serial binary adder with an interchanged role of the terminals of carry-out and sum. It can perform addition/subtraction of two Δ-Σ modulated bit-streams [5]. Thus, the main advantage of the proposed system is a simple, inexpensive, and reliable digital stereo multiplexing–demultiplexing system. As can be seen from Fig. 7.2, the proposed system consists of a number of distinct components. Thus, we will describe briefly an operation of a baseband component of the system.

7.2.1   A Transmitter Tx

Figure 7.2 shows a block diagram of a proposed stereo transmitter (Tx) and receiver (Rx). Even though any type of higher-order Δ-Σ modulator can be employed as an ADC in our simulation model, a second-order Δ-Σ modulator is used. Without loss of generality, instead of a microphone (MK1 and MK2), different transducers can be used such as a photo camera, recorded CD, or taped music. For example, a transducer can be integrated with a Δ-Σ modulator to implement a digital mike or stereo sensing system for robotic applications. Δ-Σ modulators (1 and 2) are highly oversampled where sampling frequency is much higher than a frequency of an input signal ($F_{sampling} \gg 2F_B$, where $2F_B$ is a Nyquist sampling frequency). Depending on the application, the oversampling factor R can vary ($R = F_{samp.}/2F_B$). Binary bit-streams, L and R, are added in a delta adder ΔA2 to produce a binary bit-stream

(L + R). To produce a bit-stream (L − R), the R bit-stream must be first inverted and added to a bit-stream L in ΔA1. The block diagram of a binary ΔA is shown in Chap. 2. It is an ordinary binary adder with interchanged roles of sum and carry-out terminals. According to Kouvaras [5], the sum of two delta-modulated bit-streams is defined as

$$S_n = \tfrac{1}{2}\left[Ln + Rn\right] + \text{error}. \tag{7.1}$$

The error signal consists of both quantization noise and carry-out propagation noise. This error can be minimized with proper design (a higher-order modulator) and increase of a sampling frequency. One can see that a ΔA introduces an attenuation of the sum for one-half. Depending on a sign of carry-out of a delta adder, Ln and Rn always have the same sign. Thus, Sn is always +1 or −1 [5]. If needed, one can overcome this attenuation by amplification at a receiver. Δ-Σ modulated bit-streams (L − R) and (L + R) are multiplexed using two channel multiplexers with binary address A. The output Z of a multiplexer is given by a Boolean expression

$$Z = \tilde{A} * \left(L - R\right) + A * \left(L + R\right). \tag{7.2}$$

If A = "1" (L + R) bit-stream is passed to the RF modulator and if A = "0" (Ã = "1"), then the (L − R) bit-stream is modulated. Depending on the application, a digital bit-stream can be modulated using different types of modulation such as FM, AM, ASK, etc. Radio signals can occupy different frequency bandwidths, licensed or unlicensed. Thus, a proposed stereo digital apparatus is not application limited.

7.2.2 A Receiver Rx

Stereo receiving system Rx, shown in Fig. 7.2, consists of a standard RF receiver to produce a signal Z. Digital bit-stream Z is fed into a demultiplexer whose outputs are given by Boolean expression

$$\tilde{A} * Z = \tilde{A} * \left[\tilde{A} * \left(L - R\right) + A * \left(L + R\right)\right] = \tilde{A} * \left(L - R\right) \tag{7.3}$$

and

$$A * Z = A * \left[\tilde{A} * \left(L - R\right) + A * \left(L + R\right)\right] = A * \left(L - R\right) \tag{7.4}$$

One can see if A = "1," then (L + R) is passed to both ΔA1 and ΔA2. Inverting Ã*Z one gets (−L + R). Thus, output of ΔA2 is (L + R − L + R) = 2R. Similarly, output of ΔA1 is 2L when A = 0. It is important to remember that a delta adder performs algebraic operations (not Boolean) on a delta-sigma bit-stream. Inversion means a change of sign of binary signal (+1 to −1 or −1 to +1).

7.3   Simulation Results

In Fig. 7.3, simulation results are shown when both sound signals are identical (frequency and amplitude of mike signals, MK1 and MK2). It is shown that the received signals, SPK1 and SPK2, are identical (they are inverted for clarity reasons). According to Eq. (7.1), ΔA introduces attenuation of ½ at the transmitting site. Similarly, ΔA introduces attenuation of ½ at the receiver. Thus, the total channel attenuation is ¼. During the demultiplexing process delta adder adds two identical signals; thus, the overall channel attenuation is ½. However, demodulating LP filters introduce additional attenuation, which depends on the order of a filter and its cutoff frequency.

Figure 7.4 shows an example of two input signals (MK1, MK2) of two different frequencies ($F_{MK1} = 2F_{MK2}$) and identical amplitudes. Again, one can see correctly demodulated waveforms SPK1 and SPK2. If needed, received signals SPK1 and SPK2 can be amplified.

Similarly, correct demodulation of signals is shown in Fig. 7.5, when amplitudes and frequencies of MK1 and MK2 are different.

7.4 Δ-Σ Multiplexing System-on-Chip (SoC)

Figure 7.6 shows one possible VLSI integration of a Δ-Σ arithmetic MUX units on one IC chip. A proposed IC chip consists of the following digital circuits: 4 delta adders, 4 inverters, 2-channel multiplexer, and 2-channel demultiplexer. Depending on the application, one can envision different scenarios of integration. The proposed IC system can be integrated with a delta-sigma modulator and transducer to

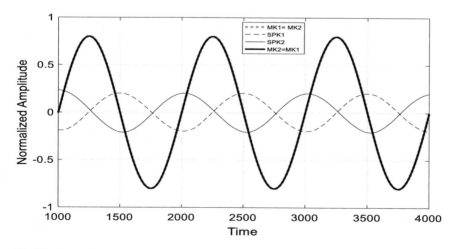

Fig. 7.3 Input signals to MK1 and MK2 have the same amplitude and frequency

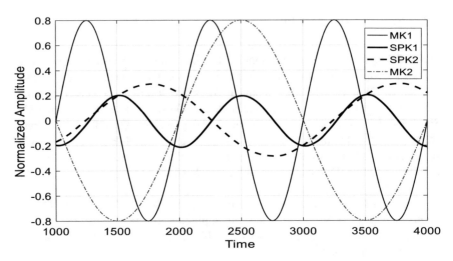

Fig. 7.4 Input signal amplitudes of MK1 and MK2 are identical, but frequencies are different

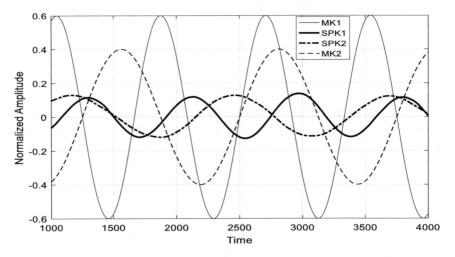

Fig. 7.5 Input signals amplitudes and frequencies of MK1 and MK2 are different

implement a system on a chip (SoC). Another option would be to multiplex 4 or more channels. Yet another possibility is to multiplex output Z with an identical output of a second IC chip. Even though the digital multiplexing technique is well-established in many areas of engineering and science, there is still space for originality. It is obvious that the possibility of direct processing of a delta-modulated bit-stream leads to a novel solution of a system-on-chip (SoC) shown in Fig. 7.6. It is also obvious that a delta-sigma multiplexed bit-stream (signal Z) can be modulated in many different ways, and transmitted wirelessly or, over other transmission media.

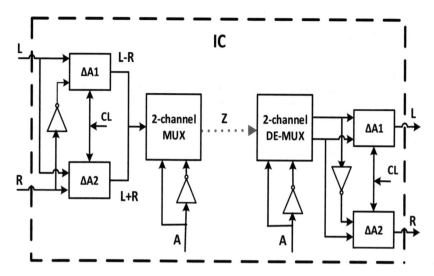

Fig. 7.6 Δ-Σ multiplexing system-on-chip [2]

Because of a serial nature of a Z bit-stream it is possible to integrate scrambler/descrambler and line encoder/decoder on the same chip as well.

7.5 Summary

Stereo methods of sound reproduction can be found in acoustic and video applications. It is common in entertainment systems such as TV, broadcast radio, cinema, videos, films, or recorded music. It can be used to navigate unmanned vehicles, robots, etc. In all applications selection of a correct distance (baseline) between right and left microphone, or cameras, is of critical importance. However, selection of correct baseline does not help if a high-resolution digital transducer (mike or camera) does not exist. A high-resolution digital delta-sigma microphone (transducer) and the proposed delta-sigma non-conventional creation of right and left channels can remedy some of the existing problems with stereo technique. Addition/subtraction of a $Δ$-$Σ$ bit-stream can lead to different arrangements of design of a system on a chip.

References

1. Tomasi, W. (1998). *Electronic communication systems* (3rd ed.). Upple Saddle River: Prentice Hall, ISBN 0-13-751439-5.
2. Zrilic, D. *U.S. Patent No.* 16/350,540.

3. Bourdopoulos, G. I., Pnevmatikakis, A., Anastassopoulos, V., & Deliyannis, T. L. (2006). *Delta-sigma modulators: modeling, design and applications*. London: Imperial College Press, ISBN 1-86094-369-1.
4. Pavan, S., Schrier, R., & Temes, G. (2017). *Understanding delta-sigma data converters*. Piscataway: IEEE Press, ISBN 978-1-119-25827-8.
5. Kouvaras, N. (1978). Operations on delta-modulated signals and their applications in the realization of digital filters. *Radio and Electronic Engineer, 48*(9), 431–438.

Chapter 8
A Δ-Σ Digital Amplitude Modulation System

8.1 Introduction

Amplitude modulation (AM) is a very well understood and developed signal processing technique. AM was the earliest modulation method used to transmit voice by radio. It was developed during the first quarter of the twentieth century. There are numerous publications on this subject, including scientific papers, books, patents, and IC data sheets. Basic principles of its operation can be found on the Internet or in any communications textbook [1, 2]. In addition to the classic analog amplitude modulation, there are several digital amplitude modulation techniques. The oldest one is on-off keying (OOK) which dates to the time of Morse telegraph transmission. Today, there are still many wireless sensors working at 433 MHz use OOK. Over the years various modulation techniques have been developed such as pulse amplitude modulation (PAM), m-ary PAM, amplitude shift keying (ASK) [2], etc. Use of the Δ-Σ modulation for RF applications is elaborated in reference [3]. It is worth mentioning that a Δ-Σ modulation is frequently referred to as a pulse density modulation [3, 4] because the density of pulses of a Δ-Σ pulse stream is proportional to the amplitude of an input signal. The proposed UC Berkeley RF pulse density system generates an amplitude-modulated waveform with up to 20 MHz envelope bandwidth and demonstrates the validity of this approach for modern communication standards [3].

To overcome the problem of a complex IF-RF digital processing, we propose a simple Δ-Σ based amplitude modulator (AM) at the transmitter and a simple asynchronous receiver. Implementation of the proposed Δ-Σ AM system is based on the dual nature of a Δ-Σ AM signal, which will be explained in the following sections.

© The Editor(s) (if applicable) and The Author(s), under exclusive license to
Springer Nature Switzerland AG 2020
D. Zrilic, *Functional Processing of Delta-Sigma Bit-Stream*,
https://doi.org/10.1007/978-3-030-47648-9_8

8.2   A Proposed Δ-Σ Amplitude Modulation System

Amplitude modulation is an increase or decrease of the carrier voltage amplitude. The carrier (high-frequency signal) carries a low-frequency message signal (the information). Thus, two signals are needed to implement modulation: one low-frequency signal and one high-frequency signal. Usually the carrier signal is a sine wave. The low-frequency signal changes amplitude, frequency, or the phase angle of the carrier. In the case of amplitude modulation, the carrier amplitude varies, with all other factors remaining constant. A newly proposed Δ-Σ amplitude modulator uses a Δ-Σ bit-stream as a carrier of a low-frequency input signal. Thus, low-frequency information (message) is contained in both the envelope and the carrier of a modulated signal. This is opposite to the classic AM transmission where information is in the envelope only and the carrier (the main power consumer) is used for demodulation purposes only. *Because* the envelope of the AM signal is prone to different kinds of noises, and its synchronous detection is more sophisticated, we propose detection of a carrier, which is a Δ-Σ pulse stream, which contains all the information about a modulating (input) signal. The block diagram of the proposed Δ-Σ AM system is shown in Fig. 8.1. Operation of the modulator (transmitter Tx) is as follows: The low-frequency analog input signal x(t) is converted into a one-bit polar bit-stream by Δ-ΣM2. The output polar bit-stream (+1, −1) of Δ-ΣM is converted into unipolar bit-stream (0, 1) by the comparator whose reference terminal is a ground, GND1. The unipolar signal $u(t)$ is then low-pass filtered (LPF) to produce a modulating signal $m(t)$. The modulating signal $m(t)$ is multiplied to produce a delta-sigma amplitude-modulated signal $S_{AM}(t)$. This signal is delivered to both an

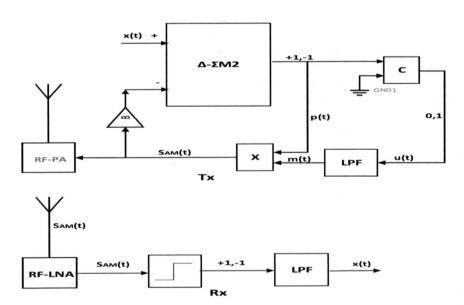

Fig. 8.1 The block diagram of the proposed Δ-Σ AM system [5]

RF transmission system and an attenuator B of a negative feedback of Δ-ΣM2. For clarity of the described process, let us write the general expression of the amplitude-modulated signal [1]

$$S_{AM}(t) = A_c \left[1 + m(t)\right] \cos \omega_c t, \tag{8.1}$$

where $[1 + m(t)]$ is a DC shifted low-frequency modulating signal, and $A_c \cos \omega_c t$ is a carrier signal. The AM waveforms and spectra of the signal (8.1) are shown in [1]. From Fig. 8.1 we can see that averaging (low-pass filtering) the unipolar signal $u(t)$, a modulating signal $m(t)$ is obtained. The signal $m(t)$ is shifted for the DC value of 0.5. Because the polar signal is a one-bit-stream (unipolar as well) representation of the input signal $x(t)$, then after a proper averaging of a unipolar signal $u(t)$ a shifted version of $x(t)$ (signal $[0.5 + m(t)]$) is obtained.

Figure 8.2 clearly illustrates this case.

Multiplying the DC shifted message signal $[0.5 + m(t)]$, shown in Fig. 8.2, with a density bit-stream of the carrier $p(t)$ we obtain an equation identical to Eq. (8.1).

$$S_{AM}(t) = \left[0.5 + m(t)\right] p(t), \tag{8.2}$$

where $p(t)$ is a polar Δ-Σ bit-stream and contains all the information about input signal $m(t)$. Thus, a Δ-Σ amplitude-modulated signal $S_{AM}(t)$ contains information in both the envelope and carrier.

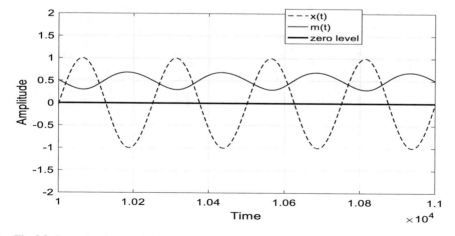

Fig. 8.2 Input signal $x(t)$ and its DC shifted version $[0.5 + m(t)]$

8.3   Simulation Results

Figure 8.3 shows both the Δ-Σ AM waveform $S_{AM}(t)$ and the demodulated signal $x(t)$. The demodulated signal, bold sinusoid in Fig.8.3, is obtained when the transmitting signal $S_{AM}(t)$ is directly connected to the sign detector (bypassing RF parts of transmitter and receiver).

Thus, passing $S_{AM}(t)$ through a sign detector circuit only zero transitions of a carrier (Δ-Σ AM bit-stream) are detected, and a polar Δ-Σ bit-stream is obtained. A Δ-Σ bit-stream is low-pass filtered to obtain the information signal $x(t)$. Thus, instead of synchronous detection of an envelope, one can asynchronously detect a carrier $p(t)$ of $S_{AM}(t)$ signal to obtain signal $x(t)$. Figure 8.4 shows the spectrum of the $S_{AM}(t)$ shown in Fig. 8.3.

In Fig. 8.1 a block diagram of a demodulator (receiver Rx) is shown as well. The operation of Rx is as follows: The AM signal $S_{AM}(t)$ is received by a low-noise radio frequency amplifier. The received signal is passed through a sign detector (zero-crossing comparator) to produce a polar signal +1, −1. Averaging this polar signal (low-pass filtering) one gets a demodulated signal $x(t)$. Figure 8.3 illustrates this case. Figure 8.5 illustrates a process of demodulation of a Δ-Σ AM signal in the presence of additive white Gaussian noise (AWGN). One can see the correct demodulation because of superior robustness of a non-positional (non-weighted) Δ-Σ bit-stream.

Figure 8.6 illustrates the case of demodulation where the index of AM modulation is regulated by the feedback attenuator/amplifier, $B = 0.5$.

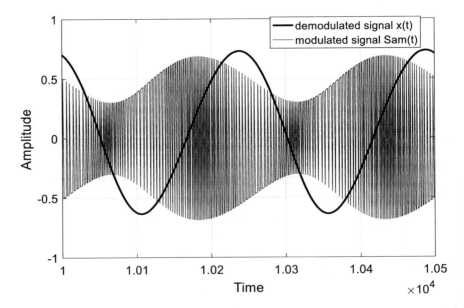

Fig. 8.3 Modulated and demodulated waveforms

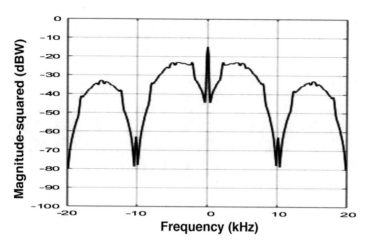

Fig. 8.4 Spectra of $S_{AM}(t)$ signal

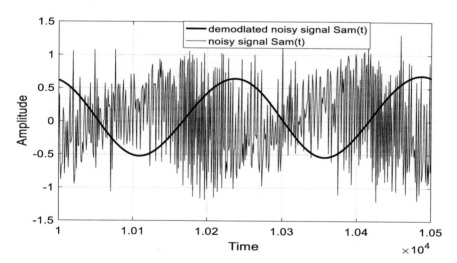

Fig. 8.5 Corrupted $S_{AM}(t)$ signal

By definition, an index of amplitude modulation is expressed as $AM_{index} = A_m/A_c$, where A_m is amplitude of a modulating signal and A_c is amplitude of a carrier signal. Figure 8.7 shows the waveforms when an amplitude of the input signal is $A_m = 0.1$, and attenuation is $B = 0.22$.

Figure 8.8 shows the case when input amplitude is $A_m = 1$. We clearly see a change in $S_{AM}(t)$ waveform as result in an A_m/A_c ratio variation.

Figure 8.4 shows the spectra of a $S_{AM}(t)$ waveform. Depending on the application, wide spectral bandwidth can interfere with neighboring channels. Thus, the bandwidth of $S_{AM}(t)$ signal must be limited. This can be achieved by reducing sampling frequency of a Δ-Σ modulator or shaping a carrier signal $p(t)$, or both.

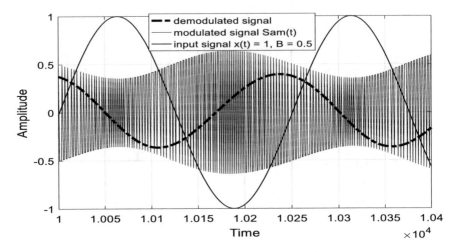

Fig. 8.6 An example of the index of AM regulation

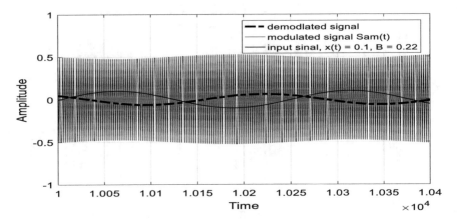

Fig. 8.7 An example of AM demodulation for low-input signal level

Figure 8.9 shows the amplitude spectra when the carrier signal $p(t)$ is slightly shaped with a low-pass filter. One can see significant reduction in a bandwidth for the same modulation index as in Fig. 8.4.

Figure 8.10 shows a shaped carrier signal $p(t)$, an input, and a demodulated output signal. We can see correct demodulation of a shaped $S_{AM}(t)$ waveform.

8.4   Summary

The idea of using a Δ-Σ modulator as a generator of an amplitude-modulated signal is based on nonlinear processing of a bit-stream in the feedback of the delta-sigma modulator. From the generated waveforms we can see that a message (information)

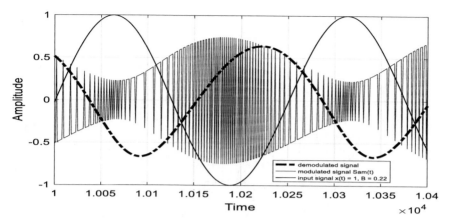

Fig. 8.8 An example of a high index of amplitude modulation

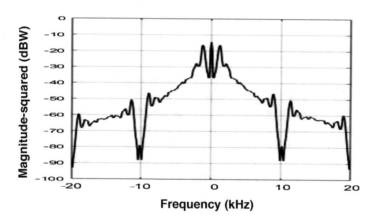

Fig. 8.9 A spectra of $S_{AM}(t)$ signal when carrier $p(t)$ is slightly shaped

is contained in both the envelope and the bit-stream of a Δ-Σ modulator. Thus, demodulation can be achieved by a zero-crossing detector and low-pass filter. Use of a low-pass filter is recommended for higher input levels of signal and a low level of channel noise. Some authors [4] consider the Δ-Σ modulator as a generator of a frequency modulated signal (pulse density is changing proportional to the input level). One can see from the generated AM waveforms that for small input levels ($V_{in} < 0.3$) and $B = 1$, the amplitude of $S_{AM}(t)$ is almost constant. Thus, the bandwidth of a Δ-Σ AM is almost the same as that of narrow-band frequency modulation (NBFM). Ideally, the FM waveform has a constant envelope. Practically, the envelope of FM contains a residual amplitude modulation that varies with time. One can see from the simulation figures an insignificant envelope variation for small input amplitudes. The Δ-Σ AM can be used for speech transmission ($F_{max} = 4$ kHz) in applications involving ambulance communications, police wireless communications, taxi wireless communications, etc. With the current state of the CMOS technology

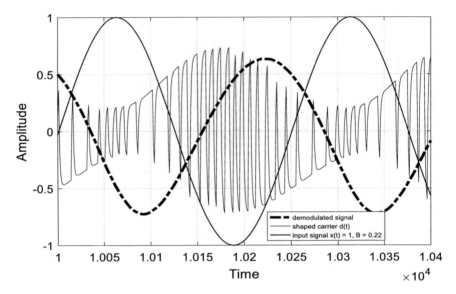

Fig. 8.10 Shaped $S_{AM}(t)$ carrier $p(t)$

oversampling factor ($R = F_{\text{sampling}}/F_{\text{in.max}}$) of 1000 or more should not be a problem
[6]. The proposed system can operate in a licensed or unlicensed frequency bandwidth. Especially, the proposed system is suitable for modulation of low-frequency
sensor signals. Our preliminary research results show the potential use of a scram-
bling technique because of the non-positional nature of Δ-Σ bits. In addition to AM,
Δ-Σ modulation has a great potential for implementation of novel types of phase and
frequency modulations. These and many other related ideas are topics for further
research.

References

1. Taub, H., & Schilling, D. (1986). *Principles of communication systems* (2nd ed.). New Delhi:
   McGraw Hill Inc. ISBN: 0-07-062955-2.
2. Tomasi, W. (2004). *Advanced electronic communication systems* (6th ed.). Upple Saddle River:
   Pearson-Prentice Hall, ISBN 0-13-045350-1.
3. Stauh, J., & Sanders, S. (2008). *Pulse-density modulation for RF applications: The radio-
   frequency power amplifier (RF-PA) as a power converter.* Berkeley, CA: University of
   California.
4. Zierhofer, C. (2008). Frequency modulation and first-order delta-sigma modulation: Signal
   representation with unit weight dirac impulses. *IEEE Signal Processing Letters, 15*, 825–828.
5. Zrilic, D. *U.S. Patent No.* 16/501,168.
6. Pavan, S., Schrier, R., & Temes, G. (2017). *Understanding delta-sigma data converters* (2nd
   ed.). Piscataway: IEEE Press, ISBN: 978-1-119-25827-8.

Chapter 9
Δ-Σ Methods for Frequency Deviation Measurement of a Known Nominal Frequency Value

9.1 Introduction

Telecommunication and electric power systems are the two manmade dynamic systems. Reliability of these systems is of great importance to the security and economic well-being of modern society. These systems extend over thousands of kilometers, and their protection and stable operation require enormous investment. To ensure stable and reliable operations many system parameters have to be monitored. One of the most important parameters in power systems is the frequency. Frequency variation can affect system operation considerably. In power systems the supply and demand must be in balance. The frequency of the power source is related to the current drawn by different loads of the grid. Both effects, overload and underload, are undesirable. Thus, it is very important that the frequency of a power system is maintained very close to its nominal frequency. There are several methods proposed in the past for frequency deviation monitoring in power systems. In reference [1] the authors proposed a frequency computation technique suitable for single- or three-phase voltage signals. This method uses a level crossing detector which yields several estimates of the frequency within one cycle. In reference [2], the authors proposed a method for the precise measurement of the difference between two low frequencies. The method is based on multiplying the two incoming frequencies by a large factor. They offer a simple and inexpensive solution for a frequency difference meter. The authors of reference [3] offer three novel techniques for frequency measurement in power networks in the presence of harmonics. All three algorithms—zero crossing, DFT method, and phase-demodulation methods—were tested and verified by simulations. A relatively simple method for measuring and display of the line frequency deviation from its nominal value is proposed in reference [4]. This counting method is based on multiplication of the reference and measured signals. The zero-crossing technique for the purpose of frequency determination of a power signal is presented in reference [5]. The Fourier algorithm is

© The Editor(s) (if applicable) and The Author(s), under exclusive license to
Springer Nature Switzerland AG 2020
D. Zrilic, *Functional Processing of Delta-Sigma Bit-Stream*,
https://doi.org/10.1007/978-3-030-47648-9_9

used for digital filtering. The verification of the proposed algorithm is done using a DAQ device in conjunction with LabVIEW. S. J. Arif [6] proposed the use of a zero-crossing detector to produce a pulse train which is combined digitally with a clock of 10 KHz and then passed through a decade counter to give the unique contribution of pulses which are encoded and displayed. This method offers a resolution of 0.5 Hz. The use of a two-arm bridge for frequency deviation measurement is proposed in reference [7]. It is an analog implementation, which uses the principle of orthogonality. This system is sensitive to input amplitude variation and additive noise as well. The use of the two-arm bridge, based on the use of Δ-Σ modulation, is proposed in reference [8]. Orthogonality is based on the use of an analog integrator. Thus, this system is sensitive to input amplitude variation. A DSP technique is proposed in reference [9]. It uses a microcomputer and data acquisition card. It offers an accurate estimate of the order of 0.02 Hz for nominal, near-nominal, and off-nominal frequencies. A ROM-based frequency deviation meter is proposed in reference [10]. It consists of two look-up tables and complex timing and control circuits. It offers near 1 mHz resolution for indoor installation with near constant temperature. The experimental circuit discussed in [11] uses PLL for multiplication of 50 or 60 Hz frequencies by 100. It has a resolution of 0.01 Hz. A programmable frequency meter for low frequencies with known nominal value is proposed in [12]. The percentage of error is 0.174% for 50 Hz. A flexible programmable circuit for generation of a radio frequency signal for transmission is proposed in US Patent No: 2006/011595 A1. This digital modulation scheme uses the well-known principle of carrier orthogonality. It is known in theory and practice as a quadrature modulation. A proposed delta-sigma transmitter consists of a quadrature digital clock generator, two delta modulators, a pair of commutators (multipliers), analog summing amplifier, band-pass filter (BPF), and antenna. Input to the system is N-bit digital word generated by a controller (micro-processor). One-bit commutator is implemented using exclusive OR (XOR) gate. The quadrature clock generator is implemented using master–slave "D" flip-flop (or a four-stage ring oscillator). A 90-degree phase difference needs only to be approximately 90-degrees. A clock frequency of a phase generator must be greater than 1 MHz (even order of GHz) to shift an information in desired radio frequency bandwidth. A transmitted signal is arbitrary, and knowledge of frequency of a source is not required. Thus, this invention is meant for the transmission of information, not for instrumentation, i.e., detection of deviation of a known nominal frequency value. The first essential component for linear processing of a Δ-Σ bit-stream is a delta adder (DA) proposed by Kouvaras [13]. A DA adder is, in fact, a binary adder with interchanged roles of sum and carry-out terminals. The second vital component, used in this chapter, is a rectifying encoder (RE) [14], which can serve as a squaring circuit operator as well [15] [US Patent 9,141,339 B2]. Implementation of the proposed methods is based on the addition and squaring operation on an orthogonal Δ-Σ modulated bit-stream. Thus, operation of these circuits is based on linear and nonlinear operations on orthogonal Δ-Σ bit-streams, in order to detect violation of the orthogonality law when a signal of known nominal frequency value changes.

9.2 A Block Diagram of the Proposed Circuits (Switch in Position A)

In Fig. 9.1 two independent methods are presented. The switch in position A shows a block diagram of a DA method.

An analog input signal of known nominal frequency is first delta-sigma modulated. Bit-stream X_n is shifted for 90° to obtain an orthogonal bit-stream X_d. Shift circuit, shown in Fig. 9.2, is implemented as a delay line which consists of serial connections of "D" flip-flops. For a known nominal frequency F_n and sampling frequency F_s, we can calculate the number of "D" flip-flops (needed to achieve 90-degree shift) of the circuit using formula: $N = F_s/4F_n = T_n/4T_s$. Thus, for $F_n = 50$ Hz and $F_s = 100$ kHz, $N = 500$. Both X_n and X_d are added in the DA to obtain bit-stream A. This bit-stream is a low-pass filtered (LPF) to obtain the analog signal $s(t)$. This analog signal is compared with a reference ground signal (zero line in Fig. 9.3) in a comparator circuit to obtain digital pulse stream C. A comparator C is a zero-crossing detector.

Demodulated orthogonal signals (X_n and X_d), output of a LP filter (bold signal $s(t)$), and output of comparator C (digital signal) are shown in Fig. 9.3. We can see, that when the demodulated signal $s(t)$ crosses zero line a digital waveform is generated. When a positive edge of digital signal C is detected, it resets the counter, and counter starts to count until the next periodic pulse edge of signal C appears again.

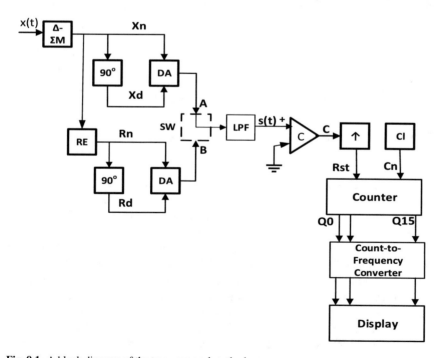

Fig. 9.1 A block diagram of the two proposed methods

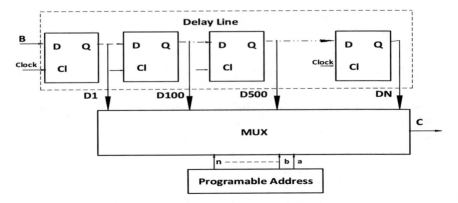

Fig. 9.2 Programmable delay line

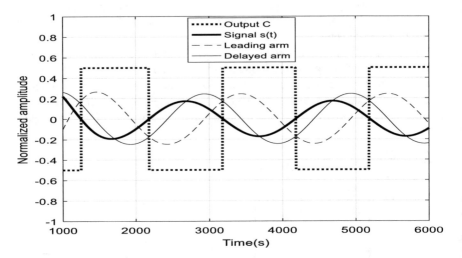

Fig. 9.3 Relevant waveforms when a switch is in position A

By counting the number of clock pulses C_n in the "window" we can calculate the frequency of the source using formula $C_{count} = T_{in}/T_s = F_s/F_{in}$, where T_{in} and T_s are periods of an input signal and a sampling signal, respectively. Digital circuits of a display can be implemented in many ways, and they are not subject to a description. A programmable 90-degree shift circuit can be implemented in diverse ways. An example of implementation using a digital delay line is shown in Fig. 9.2. It consists of a N-bit delay line, multiplexor, and address register. Operation of this circuit is well-known and its operation is described in digital design books or on the Web. For a known value of input frequency, $F_n = 50$ Hz, and a sampling frequency of $F_s = 100$ kHz, the number of "D" flip-flops, needed to achieve a 90-degree shift of signal X_n, is $N = 500$. If the desired delay line has only four distinguished outputs, then an address code word is only four-bits long. Depending on application and required resolution, we can envision different lengths of a programmable 90-degree phase shift circuit and the corresponding length of address word.

9.3   A Block Diagram of the Proposed Circuits (Switch in Position B)

Figure 9.1 (switch in position B) shows a block diagram of the (RE + DA) method. Δ-Σ bit-stream X_n is squared to obtain bit-stream R_n, which is shifted for 90° to obtain an orthogonal bit-stream R_d. Generation of the orthogonal sequence R_d can be accomplished using an identical circuit shown in Fig. 9.2. Both R_n and R_d are added in a delta adder to obtain bit-stream B, which is LP filtered to obtain an analog signal $s(t)$. This signal is compared to reference of a zero-crossing detector to obtain digital signal C whose period is two times shorter than the period of the digital signal shown in Fig. 9.3. In Fig. 9.4, operational waveforms of RE + DA method are shown: demodulated squared orthogonal sinusoidal signals R_n and R_d, output signal of LPF $s(t)$, and digital output C of comparator. Positive edges of signal C are detected to reset a counter. Because the duration of a period of binary pulses is now two times shorter, Fig. 9.4, the number of count is now $C_{count} = T_{in}/2T_s = F_s/2F_{in}$. Thus, (RE + DA) method is two times faster, because the delay line of the circuit is two times shorter than the delay line of the DA method. The rest of the circuits,

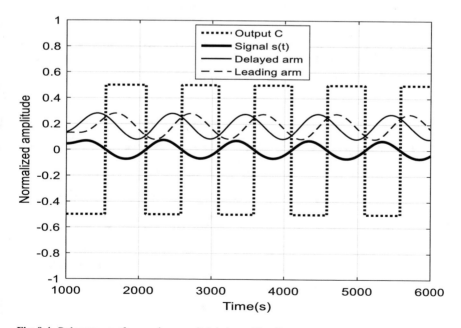

Fig. 9.4 Relevant waveforms when a switch is in position B

which serve for numerical display, are well-known in practice and the literature and can be implemented in diverse ways.

In Fig. 9.5, the rectified and squared signal of a Δ-Σ bit-stream are shown. We can see that with proper selection of LPF in Fig. 9.1, the sinusoidal signal of a double frequency (shorter period of sinusoid) can be obtained as demonstrated in Fig. 9.4.

9.4   Additional Simulation Results

In Fig. 9.6 the number of counts as a function of frequency deviation from nominal frequency (50 Hz) is shown, in the range 45–55 Hz.

We conclude that the RE + DA method is two times faster than the DA method. However, the absolute error of the DA method shows a significant advantage over the RE + DA method. As can be seen from Table 9.1 absolute error of the DA method is on average, less than 1 mHz, while for the RE + DA method absolute error is on average less than 4 mHz. In Table 9.1, measurement results are summarized for both methods, which operate at two sampling frequencies of 100 kHz and 1 MHz [16]. We observe superior performance of the DA method relative to the standard deviation.

9.5   Synchronized Frequency Deviation Meters

In Fig. 9.7, the block diagram shows the frequency synchronization difference meter. It is comprised of three or more identical Δ-Σ modulators. The reference modulator accepts stable reference signal $x_r(t)$ from a geo-position system (GPS) or

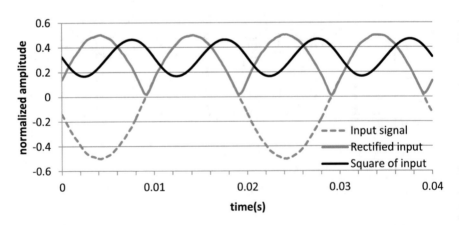

Fig. 9.5 Relevant waveforms of a Δ-Σ RE

Fig. 9.6 A number of pulse count as a function of frequency deviation

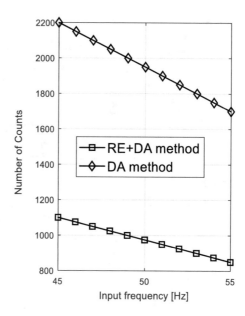

Number of Counts — Input frequency [Hz]

— RE+DA method
— DA method

Table 9.1

	$F_{sampling}$	100 kHz	1 MHz
DA method	Average	50.001	50.000
	St. dev.	0.043	0.013
	Min	49.900	49.968
	Max	50.125	50.033
RE + DA method	Average	49.996	49.998
	St. dev.	0.096	0.030
	Min	49.652	49.920
	Max	50.251	50.075

from some other stable frequency source. The rest of the modulators are in different geographical areas, which monitor input signals $x_1(t)$, $x_2(t)$, $x_3(t)$...of respective frequency generators. Either method, DA or RE + DA, can be used. In Fig. 9.7, the method of DA is employed. Outputs of comparators C1 and C3 are fed into digital comparator circuits and compared in digital comparators with reference bit-stream Cr of frequency Fr.

Digital comparator circuits can be implemented in diverse ways, such as in [15] or [16]. In Fig. 9.7, a modified version of digital comparator is adopted [16]. A digital comparator circuit comprises two 4-bit ring counters. Initially, the counters start from identical initial state "1000." One counter is clocked by Fr, and the other with F1 or F2. In Fig. 9.7, a modified version of digital comparator is adopted [16].

Figure 9.8 shows an example of synchronization of signal $x_1(t)$ with reference signal $x_r(t)$. Both signals have a nominal frequency of 50 Hz.

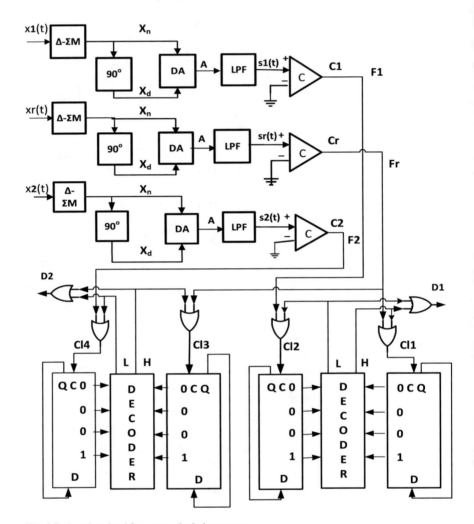

Fig. 9.7 Synchronized frequency deviation meters

The bold line in Fig. 9.8 presents a signal D1 (D1 = 0) at output of a digital comparator. Figure 9.9 shows the case when frequency deviation of a signal $x_1(t)$ is 0.1 Hz (input frequency 49.9 Hz).

We can see the appearance of pulse stream when a nominal reference and a measured frequency are discrepant. It is easy to conclude that a digital pulse stream D1 (or D2) can be used for monitoring and synchronization of multiple frequency sources (power generators for example). Signal D1 can be averaged and fed back for control and synchronization purposes.

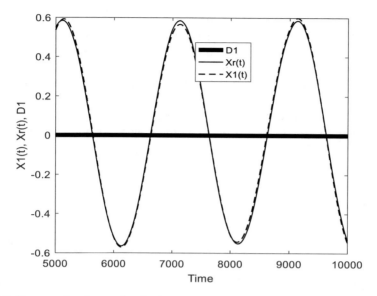

Fig. 9.8 The case of perfect synchronization

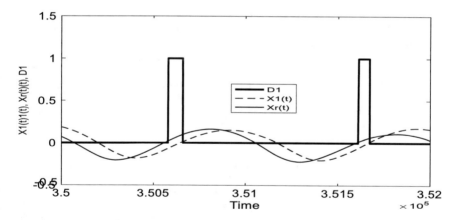

Fig. 9.9 The case when error is detected

9.6   Summary

Δ-Σ modulation is a high-resolution, low-power consuming, and inexpensive analog-to-digital converter (ADC). It is an ideal ADC for interfacing purposes at low-frequency and low-level sensing signals. It has a wide range of applications such as in instrumentation, industrial electronics, sensors, communication, and control. The newly proposed methods, based on direct processing of a Δ-Σ bit-stream, can be used to measure frequency deviation of the source with a known nominal frequency in applications such as power generators, EKGs, engine health monitoring, etc. In addition, the proposed methods can be used for synchronization or control purposes of dislocated frequency generators. The proposed methods are based on non-conventional processing (linear and nonlinear) of an orthogonal bit-stream of one delta modulator. Any change of frequency of known value causes the violation of orthogonality law. A violation signal is detected at the output of the delta adder (DA) and further processed to obtain a quantitative value of deviation of frequency of known nominal value. Detected violation pulses can be used for control and synchronization purposes or transmitted to control center via a dedicated communication line.

References

1. Nguyen, C. T., & Drinivansan, K. (1984). A new technique for rapid tracking of frequency deviation based on level crossing. *IEEE Transactions on Power Apparatus and Systems, PAS-103*(8), 2230–2236.
2. Kaspris, T., Loulgaris, N., & Halkia, C. (1985). A method for the precise measurement of the difference between two low frequencies. *IEEE Transactions on Instrumentation and Measurement, IM-34*(1), 95–96.
3. Begovi, M., Djuric, P., Dunlap, S., & Phadke, A. (1993). Frequency tracking in power networks in the presence of harmonics. *IEEE Transactions on Power Delivery, 8*(2), 480–486.
4. Dwivedi, J., Shukla, M., Verma, K., & Singh, R. (2010). A novel technique for indication of power frequency deviation in electrical systems. In V. Das, J. Stephen, N. Thankachan, et al. (Eds.), *International Conference on Power Electronics and Instrumentation Engineering, CCIS 102* (pp. 80–82). Berlin: Springer.
5. Chen, Y., & Chien, T. (2015). A simple approach for power signal frequency determination on virtual internet platform. *Applied Mathematics & Information Sciences, 9*(1L), 65–71.
6. Arif, S. J. (2016). A simple and efficient method for accurate measurement and control of power frequency deviation. *International Journal on Electrical and Computer, Energetic, Electronic and Engineering, 10*(6), 2916.
7. Ahmad, M. (1988). Power system frequency deviation measurement using an electronic bridge. *IEEE Transactions on instrumentation and measurement, 37*(1), 147–148.
8. Zrilic, D., & Pjevalica, N. (2004). Frequency deviation measurement based on two-arm Δ-Σ modulated bridge. *IEEE Transactions on Instrumentation and Measurement, 53*(2), 293.
9. Sidhu, T. S. (1999). Accurate measurement of power system frequency using a digital signal processing technique. *IEEE Transactions on Instrumentation and Measurement, 48*(1), 75–81.
10. Banta, L. E., & Xia, Y. (1990). A ROM-based high accuracy line frequency and line deviation meter. *IEEE Transactions on Instrumentation and Measurement, 39*(3), 535–539.
11. Nemat, A. (1990). A high-resolution digital frequency meter for low frequencies. *IEEE Transactions on Instrumentation and Measurement, 39*(4), 667.
12. Irshid, M. I., Shahab, W. A., & El-Asir, B. R. (1991). A simple programmable frequency meter for low frequencies with known nominal values. *IEEE Transactions on Instrumentation and Measurement, 40*(3), 640–642.
13. Kouvaras, N. (1978). Operations on delta-modulated signals and their applications in the realization of digital filters. *Radio and Electrical Engineer, 48*(9), 431–438.
14. Zrilic, D., Petrovic, G., & Tang, W. (2017). Novel solution of a delta-sigma-based rectifying encoder. *IEEE Transactions on Circuits and Systems II: Express Briefs, 64*(10), 1242.
15. Zrilic, D. G., "Method and apparatus for full-wave rectification of delta-sigma modulated signals," *U.S. Patent No.* 9,525,430 B1.
16. Zrilic, D., & Pjevalica, N., Unpublished joint research results.

Chapter 10
Δ-Σ Automatic-Gain Controller

10.1 Introduction

Automatic-gain control (AGC) is a closed-loop feedback regulating circuit. It is used in many systems to maintain a nearly constant signal amplitude at its output despite variation of the signal amplitude at the input. Its role is to reduce the amplitude dynamic range so that circuits following the AGC can handle a reduced dynamic range. In 1925, Harold A. Wheeler invented and patented automatic volume control (AVC). Karl Küpfmüller published an analysis of AGC systems in 1928. By the early 1930s most new commercial broadcast receivers included AVC [1]. AM radio has used AGC since then. It is used in most radio receivers to equalize the average volume (loudness) of different radio stations due to differences in the received signal strength, as well as variations in a single station's radio signal due to fading. Related applications of AGC are in radar systems, telephone recording, audio/video recording, biological sensory systems, etc. Correct operation of mobile telephone systems is greatly dependent on AGC due to environmental factors such as seasonal change in vegetation, fluctuation of users, overpower of a neighboring base station, etc. A block diagram of a typical AGC is shown in Fig. 10.1. In brief, we will describe its operation as presented in [2]. As can be seen, the output of VGA, signal V1, is fed back to the detector of the amplitude change and to the next device in the signal-processing chain as well. If needed, signal V1 can be amplified. A detector block can be implemented as an envelope (or rectifier), square law, true RMS, or logarithmic. A detected signal V2 is smoothed (low-pass filtered) to produce a nearly DC signal. This signal is compared with a reference signal VR. The result of the comparison is used to generate the control voltage Vc to adjust the gain of the VGA. There are many publications, books, patents, VGA circuit solutions, and Internet sites dealing with theory, implementation, and applications of AGC devices. There are IC chips on the market as well. For instance, Analog Devices Inc. application note describes a low-frequency AGC circuit (AD8336) used in audio

© The Editor(s) (if applicable) and The Author(s), under exclusive license to Springer Nature Switzerland AG 2020
D. Zrilic, *Functional Processing of Delta-Sigma Bit-Stream*, https://doi.org/10.1007/978-3-030-47648-9_10

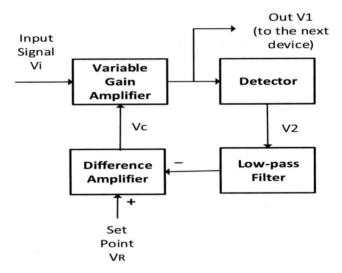

Fig. 10.1 Basic architecture for an AGC system

and power equipment [3]. A block diagram of a delta-sigma based automatic-gain control circuit is presented in reference [4]. However, rectification of analog input signal is done off delta-sigma IC chip, with the use of analog components.

A compressor circuit presented in Chap. 6 can operate as an AGC circuit as well. With slight modification of a compressor circuit, we present a novel AGC circuit based on nonlinear processing of delta-sigma bit-stream, where rectification is performed directly on delta-sigma bit-stream. Depending on the application, capacitor C could be an eventual external component of IC chip.

10.2   Proposed Δ-Σ AGC

In Fig. 10.2, a novel architecture of AGC is presented. A second-order Δ-Σ modulator is used as an ADC, but without loss of generality, the first-order or any higher-order Δ-Σ modulator can be used. In brief, operation of the proposed AGC is as follows.

An input signal is converted into a polar bit-stream $D\varepsilon\{-1 + 1\}$ by means of a Δ-Σ modulator whose negative input terminal is controlled by signal V_c. The output D of a Δ-Σ modulator is low-pass filtered and amplified (if needed) to generate an adequate level V_{out}. At the same time, a polar signal D is converted to a unipolar Δ-Σ bit-stream, and its signal V_n is delivered to a rectifying encoder (RE) [5]. A rectified unipolar bit-stream is low-pass filtered to produce a slow changing signal $\sim V_{dc}$ (nearly DC signal).

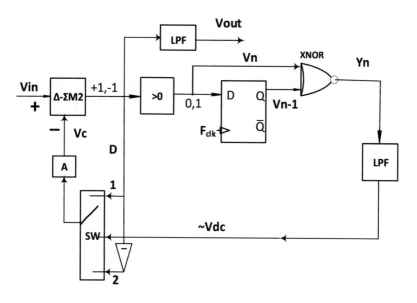

Fig. 10.2 Δ-Σ architecture of the proposed VGC

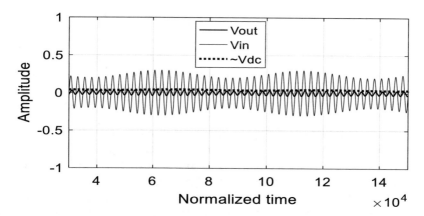

Fig. 10.3 Waveform relations of AGC circuit ($A = 4$)

10.3 Simulation Results

The relation elation between signals V_{in}, V_{out}, and $\sim V_{dc}$ is shown in Fig. 10.3. Amplitude of the input modulating AM signal is $V_{in} = 0.1$, and sampling frequency of delta Δ-ΣM2 is $F_s = 50$ kHz. In this case, amplification in the feedback of a delta-sigma modulator is $A = 4$. We see significant reduction in amplitude fluctuation of a signal Vout at the output of AGC.

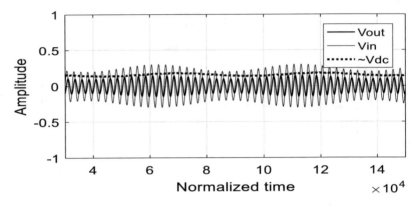

Fig. 10.4 Waveform relations of AGC circuit without amplification ($A = 1$)

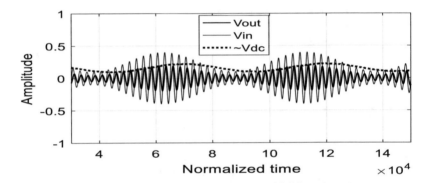

Fig. 10.5 Waveforms relations of an AGC circuit for highly fluctuating input signal ($A = 1$)

However, depending on the application, a feedback amplifier can be excluded. Figure 10.4 shows this case ($A = 1$) for the same input signal and sampling frequency as in Fig. 10.3. We cannot see a significant difference in AGC waveforms when a change of level of input signal is relatively slow (not abrupt).

Without an amplifier for a higher level of AM modulating signal, $V_{in} = 0.3$, of input signal, the AGC circuit exhibits significant fluctuation. This case is shown in Fig. 10.5.

Thus, for a high-fluctuating input signal, use of a feedback amplifier is recommended. Figure 10.6 shows the same waveforms as in Fig. 10.5 when amplification $A = 4$. We can see that the ratio of A_{max}/A_{min} (dynamic) of input signal is $0.4/0.1 = 4$, while the ratio of AGC signal is $0.05/0.025 = 2$.

In Fig. 10.7, enlarged waveforms of Fig. 10.6 are presented. We can now estimate the ratio of the A_{max}/A_{min} (dynamic) of input signal is $0.4/0.1 = 4$, while the ratio of AGC signal is $0.05/0.025 = 2$.

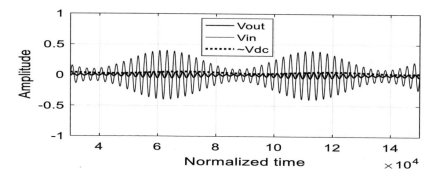

Fig. 10.6 Waveform relations of AGC circuit with amplifier, $A = 4$

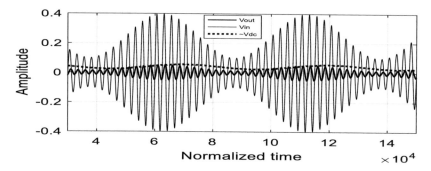

Fig. 10.7 Enlarged signals of Fig. 10.6

10.4 Summary

Existing implementations of VGAs were based on analog multipliers and a substantial number of discrete passive elements. These solutions are inherently sensitive to the components parameter mismatch. Even a very small drift of parameters in discrete passive components can cause problems. Some of the VGA components may require multiple supplies as well. A VGC proposed by Dias [4] is based on delta-sigma modulation. A rectification of an analog input signal is done with an off-IC chip analog components. To overcome these problems, we propose robust VGC building components such as a delta-sigma modulator, on-chip digital rectifier of delta-sigma bit-stream, switching mode amplifier, and a simple low-pass filter. A novel VGC architecture is not application limited and it reduces the number of off-chip components. It operates at a single supply voltage and is suitable for IC design. Depending on the application, the only an external component could be a capacitor of a low-pass filter. For complete validation of the proposed VGC system, it is necessary to investigate attack/recovery time as a function of a sampling frequency of a delta-sigma modulator, linearity of a gain as a function of a signal input level, etc.

References

1. Wikipedia, Automatic Gain Control.
2. Rosu, I., Automatic Gain Control (AGC) in Receivers, *YO3DAC/VA3IUL*. Retrieved from http://www.qsl.net/va3iul/
3. *Analog Devices Inc*. AD8336 IC, general purpose amplifier.
4. da Fonte Dias, V. (1996). Signal processing in the delta-sigma domain. *Microelectronics Journal, 26*(6), 543–562.
5. Zrilic, D., Petrovic, G., & Tang, W. (2017). Novel solution of a delta-sigma-based rectifying encoder. *IEEE Transactions on Circuits and Systems II: Express Briefs, 64*(10), 1242.

Chapter 11
Δ-Σ Integrator and Differentiator

11.1 Introduction

Digital integrators and differentiators are an integral part of many systems like control, communications, audio, medical, and DSP in general. Both digital integrators and differentiators can be classified as finite impulse response (FIR) or infinite impulse response (IIR) filters. IIR digital solutions are preferred because their implementation is simpler, and they have much better magnitude response. However, their phase characteristics are not linear in comparison with FIR filters, which can cause problems in some applications. The frequency response of an ideal integrator is

$$H_I\left(e^{j\omega}\right) = 1/j\omega. \tag{11.1}$$

Practical digital integrators are designed to have a frequency characteristic approximating expression (11.1). The frequency response of an ideal discrete-time differentiator is given by

$$G_D\left(e^{j\omega}\right) = j\omega. \tag{11.2}$$

The magnitude of the frequency response of the ideal integrator falls between the magnitude of the rectangular and the trapezoidal integrators. Approximating the ideal integrator as a weighted sum of the rectangular and the trapezoidal integrators, Al Alaoui proposed a novel integrator and differentiator [1]. Time-domain input–output relations of the first-order rectangular IIR digital integrator are

$$y[n] = y[n-1] + T * x[n-1], \tag{11.3}$$

where T is the sampling period. Its transfer function is given by

© The Editor(s) (if applicable) and The Author(s), under exclusive license to
Springer Nature Switzerland AG 2020

D. Zrilic, *Functional Processing of Delta-Sigma Bit-Stream*,
https://doi.org/10.1007/978-3-030-47648-9_11

$$H_R(z) = T * (1/(z-1)). \tag{11.4}$$

Time-domain input–output relation of a trapezoidal integrator is

$$y[n] = y[n-1] + T/2 * [x[n] - x[n-1]]. \tag{11.5}$$

Its transfer function is

$$H_T(z) = T/2 * [(z+1)/(z-1)]. \tag{11.6}$$

According to Al Alaoui, the weighted sum of (11.4) and (11.6) is

$$H_N(z) = (3/4)[T/(z-1)] + (1/4)(T/2)(z+1)/(z-1)$$
$$= (T/8)[(z+7)/(z-1)]. \tag{11.7}$$

Modifying expression (11.7) a non-minimum phase transfer function of the digital integrator is obtained [1]:

$$H(z) = 7 * (T/8) * (z+1/7)/(z-1). \tag{11.8}$$

By inversion of the transfer function of the integrator, an Al Alaoui differentiator is obtained:

$$G(z) = 8(z-1)/7T(z+1/7). \tag{11.9}$$

Both the integrator and differentiator are the first order.

Keeping in mind the above approach, the objective of this chapter is to propose a novel delta-sigma digital integrator and differentiator whose operations are based on linear arithmetic operations on delta-sigma bit-stream.

11.2   New Delta Adder

A delta-sigma integrator (Δ-ΣI) consists of the serial connection of a FIR and IIR filter. Using delta adder as a building block (presented in Chap. 2), we can obtain the time-domain input–output relation of the integrator:

$$Y[n] = \{[x[n] + x[n-1]]/2 + y[n-1]\}/2. \tag{11.10}$$

Taking z-transform of Eq. (11.10), we obtain the following transfer function of the proposed Δ-ΣI:

$$HΔ-ΣI(z)=(1+z^{-1})/2(2-z^{-1})=0.5[(z+1)/(2z-1)]. \quad (11.11)$$

For $\omega = 0$, an ideal integrator has a singularity point (pol), while $Δ$-$ΣI$ has a pol at $z = 1/2$ ($z = e^{j\omega}$).

To overcome this problem, we propose a novel delta adder (NΔA), which consists of 4 logic gates and D flip-flop. A Boolean logic equation for the sum of two delta-sigma modulated bit-streams is

$$S_n = N\{X_nY_n + S_{n-1}(X_n \bmod 2Y_n)\}. \quad (11.12)$$

A logic diagram of implementation is shown in Fig. 11.1.

Its logic truth table and equivalent algebraic truth table are given below. Substituting a logic "0" by "−1" we can obtain the algebraic equation (11.13):

Logic Truth Table

	X_n	Y_n	S_{n-1}	S_n
0	0	0	0	1
1	1	0	0	1
2	0	1	0	1
3	1	1	0	0
4	0	0	1	1
5	1	0	1	0
6	0	1	1	0
7	1	1	1	0

Algebraic Truth Table

	X_n	Y_n	S_{n-1}	S_{n_a}
0	−1	−1	−1	−2
1	1	−1	−1	−2
2	−1	1	−1	−2
3	1	1	−1	2
4	−1	−1	1	−2
5	1	−1	1	2
6	−1	1	1	2
7	1	1	1	2

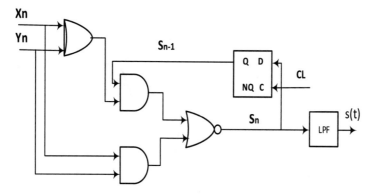

Fig. 11.1 Logic diagram of a new delta adder (NΔA)

$$S_{n\_a} = X_n + Y_n + \left(1 - X_n Y_n\right) S_{n-1}, \qquad (11.13)$$

where S_{n_a} stands for algebraic sum. We see that the algebraic truth table above is identical to the algebraic truth table No. 1 of a Kouvaras' delta adder (Chap. 2). The only difference is in the column for the algebraic sum, where "−1" is multiplied by "2." Thus, a newly proposed adder produces a signal without attenuation of one-half.

11.2.1   Simulation Results

To verify the operation of the proposed NΔA, simulations are performed using Simulink software. Figure 11.2 shows the case of addition of two sinusoidal signals with the same amplitude and frequency. The input signal is presented with a solid line and a signal of the sum with the dotted line. We see that a signal of the sum is not attenuated as was the case with a classic delta adder [2, 3].

In Fig. 11.3, two sinusoidal signals of different frequencies (f2 = f1/2) and the same amplitudes are added. In both figures the oversampling factor is $R = Fs/f_{in} = 5000$, and as a demodulator LP Butterworth filter of the second-order is used.

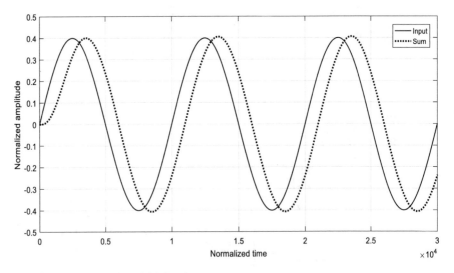

Fig. 11.2 Sum of two sinusoidal signals

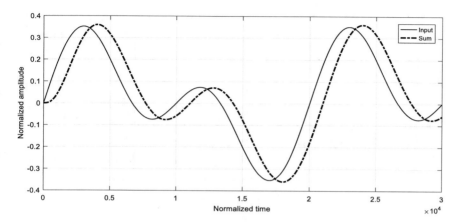

Fig. 11.3 Sum of two signals of different frequencies

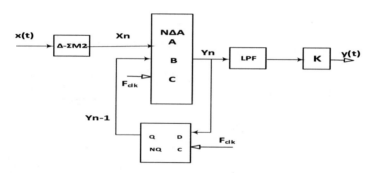

Fig. 11.4 Block diagram of the proposed Δ-ΣI

11.3 Δ-Σ Integrator

In Fig. 11.4, the simplest delta-sigma integrator is presented. Its implementation is based on the discrete-time equation

$$Y_n = X_n + Y_{n-1}. \tag{11.13}$$

Taking z-transform of Eq. (11.13) we obtain a transfer function of the integrator

$$TI(z) = 1/\left[1 - z^{-1}\right]. \tag{11.14}$$

Verification of the operation of the proposed integrator is performed by extensive simulation using Simulink software. In Figs. 11.5 and 11.8, time-domain wave-

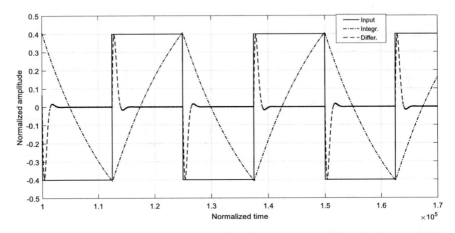

Fig. 11.5 Relevant waveforms of the proposed integrator and differentiator

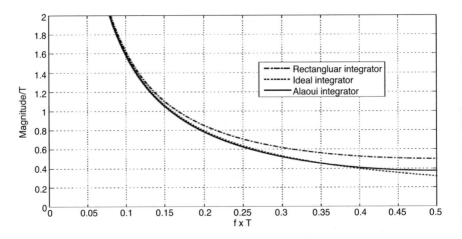

Fig. 11.6 Magnitude response of the proposed Δ-Σ integrator

forms are shown for both the integrator and differentiator for abrupt input signals. In Fig. 11.6, the magnitude response of the proposed integrator is presented. On the same diagram, for comparison purposes, ideal and Al Alaoui filter magnitude responses are plotted. Keeping in mind a simple realization of the first-order delta-sigma integrator, we can conclude that the performances of the new integrator are not optimal, but in some applications satisfying.

11.4 Δ-Σ Differentiator

A newly proposed delta-sigma differentiator (Δ-ΣD) circuit block diagram is shown in Fig. 11.7. Its implementation is based on the use of a NΔA as well. Its operation is described by a difference equation

$$Y_n = X_n - X_{n-1}. \tag{11.15}$$

Taking z-transform of Eq. (11.15) we obtain a transfer function of the differentiator

$$TD(z) = 1 - z^{-1}. \tag{11.16}$$

In Figs. 11.5 and 11.8, time-domain waveforms are shown for both differentiator and integrator, when input to the Δ-Σ modulator is an abrupt signal (square wave or sawtooth). The magnitude response of a rectangular differentiator, which is implemented using NΔA, is shown in Fig. 11.9.

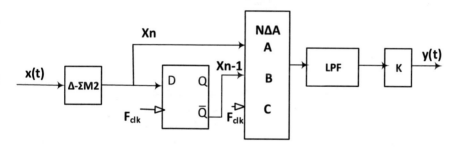

Fig. 11.7 Block diagram of a proposed Δ-ΣD

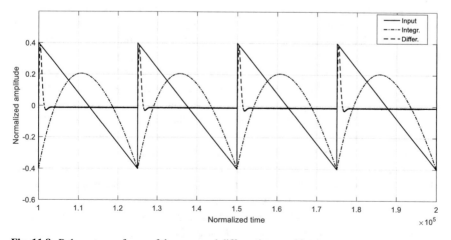

Fig. 11.8 Relevant waveforms of the proposed differentiator and integrator

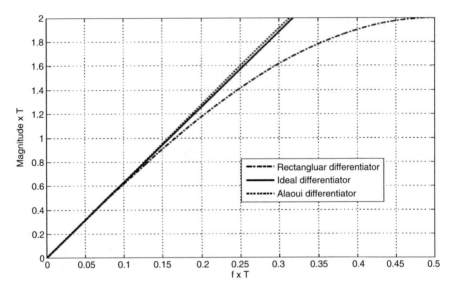

Fig. 11.9 Magnitude response of the proposed Δ-Σ differentiator

11.5   Conclusion

A novel type of delta adder is presented. Its implementation is based on the use of a
modified version of the binary half-adder. It consists of 4 logic gates and D flip-flop.
This adder does not introduce attenuation, as was the case of delta adders proposed
in references [2, 3]. At higher input levels it introduces a slight amplification. Using
this adder, we demonstrated its possible application in implementation of a simple
and inexpensive rectangular integrator and differentiator. Simulation results have
shown the correctness of the operation even though there is some deviation from
ideal transfer function at higher frequencies of input signal. Thus, there is room for
further research and eventual optimization, in particular, optimization of delta adder
for all input levels to the Δ-Σ modulator. Simulation waveforms show in Fig. 11.2 and
Fig. 11.3 are obtained using a discrete model of a second-order Δ-Σ modulator, which
has a feed-forward coefficient equal to one and a feed-back coefficient equal to 0.5.
For example, it is necessary to examine the performance of higher-order Δ-Σ integrators and differentiators or approximating the ideal integrator as a weighted sum of the
proposed rectangular delta-sigma and the trapezoidal integrators (Al Alaoui's
approach). In addition, it is necessary to examine the operation of a proposed circuit
as a function of the oversampling factor of the delta-sigma modulator and the order
of low-pass demodulating filter. It is important to point out that the digital integrator's
accuracy performance is not solely a function of its frequency-domain response. Its
performance is also profoundly dependent upon the input signal's bandwidth, mono-
tonicity, sampling, number of derivative polarity changes, etc. [4].

References

1. Al-Alaoui, M. (1993). Novel digital integrator and differentiator. *Electronics Letters, 29*(4), 376–378.
2. Kouvaras, N. (1978). Operations on delta-modulated signals and their applications in the realization of digital filters. *Radio and Electrical Engineer, 48*(9), 431–438.
3. Fujisaka, H., Kurata, R., Sakamoto, M., & Morisue, M. (2002). Bit-stream signal processing and its applications to communication systems. *IEE Proceedings-Circuits, Devices and Systems, 149*(3), 159–166.
4. Lyons, R. (2019, September 24), The risk in using frequency domain curves to evaluate digital integrator performance, *DSP Related.com*. Retrieved from https://www.dsprelated.com/show-article/1299.php

Conclusion

Delta-sigma modulators are one-bit serial analog-to-digital converters distinguished by their high-resolution and simple design in comparison with n-bit analog-to-digital converters. The problem of direct arithmetic operation on a serial delta-sigma bit-stream must be solved by design of special circuits. Some of the advantages of delta-sigma ADS are:

- the possibility of direct linear, nonlinear, and mix analog/digital operations on its bit-stream,
- the simplicity of the processing circuits, and
- the possibility of obtaining results in both digital and analog form.

Direct processing of a low-pass delta-sigma bit-stream possesses further advantages in solving each specific problem. For instance, in environmental sensing systems, wearable devices, automatically controlling slow changing processes, etc.

Advancement of nano-technologies will enable the design of simple, low-power consuming linear and nonlinear signal processing functional units including adder/subtractor, rectifier, multipliers by a constant, multiplication of two signals, integrators, as well as function generators.

Thus, non-conventional DSP, based on delta-sigma modulation ADC, makes possible the development of new approaches in the design and connection of individual functional units in different physical processes.

We hope that the newly proposed circuits introduced in this monograph will trigger increased interest in the non-conventional signal processing field.

© The Editor(s) (if applicable) and The Author(s), under exclusive license to
Springer Nature Switzerland AG 2020
D. Zrilic, *Functional Processing of Delta-Sigma Bit-Stream*,
https://doi.org/10.1007/978-3-030-47648-9

Index

© The Editor(s) (if applicable) and The Author(s), under exclusive license to
Springer Nature Switzerland AG 2020
D. Zrilic, *Functional Processing of Delta-Sigma Bit-Stream*,
https://doi.org/10.1007/978-3-030-47648-9

Printed in the United States
by Baker & Taylor Publisher Services